D1497905

VITAL CIRCUITS

VITAL

STEVEN VOGEL

Department of Zoology
Duke University

ILLUSTRATED BY
Rosemary Anne Calvert

New York   Oxford
OXFORD UNIVERSITY PRESS
1992

CUMBERLAND COUNTY COLLEGE LIBRARY
P.O. BOX 517
VINELAND, N.J. 08360

CIRCUITS

On Pumps, Pipes, and the
Workings of Circulatory Systems

QP
102
V65
1991

92-140

Oxford University Press

Oxford   New York   Toronto
Delhi   Bombay   Calcutta   Madras   Karachi
Petaling Jaya   Singapore   Hong Kong   Tokyo
Nairobi   Dar es Salaam   Cape Town
Melbourne   Auckland

and associated companies in
Berlin   Ibadan

Copyright © 1992 by Steven Vogel

Published by Oxford University Press, Inc.,
200 Madison Avenue, New York, New York 10016

Oxford is a registered trademark of Oxford University Press

All rights reserved. No part of this publication may be reproduced,
stored in a retrieval system, or transmitted, in any form or by any means,
electronic, mechanical, photocopying, recording, or otherwise,
without the prior permission of Oxford University Press.

Library of Congress Cataloging-in-Publication Data
Vogel, Steven, 1940–
Vital circuits : on pumps, pipes, and the workings of
circulatory systems / Steven Vogel ; illustrated by Rosemary Anne
Calvert.
p. cm.   Includes bibliographical references and index.
ISBN 0-19-507155-7
1. Cardiovascular system—Physiology.   I. Title.
QP102.V65   1991
612.1—dc20   91-23954

1 3 5 7 9 8 6 4 2

Printed in the United States of America
on acid-free paper

To my mother, Jeanette Vogel,
and my wife, Jane Vogel,
the two women closest to my heart.

Preface

The good part of writing a book on circulation is that one can build on everyday objects and experiences. Little recourse is necessary to such things as molecules whose existence, at least for most of us, requires something of a leap of faith. The bad part comes when one has to pick a title. Using "blood" promises a murder mystery; using "heart" suggests a Gothic romance.

In this book several pieces of my life come together. For one thing, I'm an odd sort of biologist who looks at problems involving the conjunction of biological function and physical fluid mechanics. For another, I'm a scientist with at least the normal introspective curiosity about what we do and how we do it. Beyond that, I'm a teacher and writer, distressed at how little people know about how science works and at the pervasive perversions of the process in the popular press. And I live in the aftermath of a heart attack, a patient at a cardiac rehabilitation clinic who interacts with participants on both sides of its ledger. Any piece alone might have provided sufficient impetus for a book; in combination, as you can see, their provocation has proven irresistible.

While I certainly worry a lot about circulatory systems (mainly my own, of course), I must admit at the outset that I've never done any specific investigation on the subject. I was once the second author of a short paper on circulation, but that was the serendipitous outcome of a lecture created for an introductory biology course. Still, the combination of training in animal physiology and experience with fluid mechanics makes accessible a large body of scientific and medical literature; that literature has provided the basic material for the present book.

I have tried to provide what I regard as a proper biological context for all one hears about cardiovascular disease. To me that context is a most engaging story—with main and subsidiary characters, with a decent plot, with fascinating side issues, with applications to nonliving technology, with historical implications, and with glimpses of the lurching progress of science. I have tried not to dwell on specifically medical concerns, the problems of systems out of tune. Anyone who totally ignores such matters is clearly living in a fool's paradise, but lots of relevant material about our troubles is readily available. What hasn't been told is the good side of the story, the nonpathological side, the biological epic rather than the human tragedy. About 30 years ago, when I decided to go to graduate and not to medical school, I chose to focus on that epic.

In particular, the story hasn't been told for a reader uninitiated into either the arcana of medicine and pharmacology or of biology and evolution. In a way, this book is an experiment—can I use my experience in teaching both neophyte biologists and nonscientists to write about an unrelievedly scientific subject? To put a finer point on it, can I do so without the more or less captive audiences upon whom I'm ordinarily imposed? It's an experiment of another sort as well—can I use our concern with cardiovascular disease to draw attention to science itself, at once the most ordinary and the most fascinating of human cultural activities?

In almost every popular account, doing science is either explicitly or implicitly justified by its relevance to the material welfare of humans. I have very strong feelings that such justification is inadequate for science as a whole and misleading with regard to the motivations of scientists. To me, the place of science transcends practicality or technology—it's a cultural phenomenon, the greatest of all products of human spirit and intellect. It ought to enrich our lives by providing a more powerful view of ourselves and our world, playing just the same role as do the arts. To play that role effectively, it must be accessible to an audience wider than some specifically anointed elite; but in large measure it's not so accessible, although it may have been in an earlier era.

The criticism of inaccessibility that's often leveled at elements of contemporary art, music, and literature applies far more forcefully to science. The remedy, at least for science, is a combination (if the two are separable anyway) of education and communication, which is what I'm attempting here. Science has great beauty, but mostly it's the beauty of a great piece of music that exists only as a score and is never given public performance.

I did say "education," and what the reader will find here has a mildly pedagogical character. To some extent that's merely one of the author's professional liabilities. Mainly, though, it's deliberate—I want to draw the reader in, rather than just provide information and anecdotes about a world to which ordinary people cannot hope to achieve proper admission. The result is a book of some density, the latter only disguised, not transformed or diluted, by a little playfulness of prose. Whether a reader takes seriously such things as the units in which pressure is expressed is a matter of personal taste and intention—but at least the material is here, ignorable if that is one's pleasure. To mention one can be quick and slick; to explain takes a little more space, more sensitivity, and some sacrifice of speed of gratification.

Even beyond the deliberate focus on normal function, two omissions may strike some as conspicuous "dogs that didn't bark." First, electrical activity of hearts in general and electrocardiograms in particular get almost no attention. That's quite deliberate. An intuitively satisfying introduction to bioelectricity is quite beyond what I can do in a book mainly about something else. In addition an electrocardiogram is essentially a correlative tool. That is, while we have an idea of what's normal and what pathologies are signalled by various deviations, the normal pattern reveals little of basic functional significance. Second, chemistry in general and more specifically biochemistry, endocrinology, and neuropharmacology get little attention. Again, one can't do everything, and I have quite deliberately emphasized topics for which I thought I could provide the reader with much more than ex-cathedra pronouncements.

The fastidious fact-by-fact documentation enjoined upon us scientists from our intellectual infancies has been, quite simply, omitted. My rationale (or excuse) is that most of what's here is common knowledge—this book contains no novel synthesis or emergent hypothesis but is, as mentioned, an exercise in making material accessible to people without special background or terminological facility. While the sources I've consulted are listed in the final bibliography, I ought to mention a few that have proven particularly useful. For the fluid mechanics, Caro et al. (1978)—complex but authoritative and unambiguous. For basic circulatory physiology, Folkow and Neil (1971), Burton (1972), and Milnor (1990). For what clinical correlates I cared about, the fifteenth edition of the Merck Manual (1987). For perspective and basic guidance on nonhuman systems, the fourth edition of the comparative physiology text of my colleague, Knut Schmidt-Nielsen (1990). This last is probably the only one of these accessible to the nonspecialist.

I have always found piecemeal reading of gestating books something less than a joyous activity. As a result I am at once puzzled and exceedingly grateful that friends and/or relatives express any willingness to have a look at my stuff. Their kindness in taking on such a chore is much appreciated, even if I on occasion wish a little less kindness and forbearance in their criticism. All or major portions of the present manuscript have been read and remarked upon by Hugh Crenshaw, Lars Ekelund, Matthew Healy, Michael LaBarbera, Alexander Motten, Jane Vogel, Max Vogel, and Stephen Wainwright. In addition, William Kier, Ann Pabst, Michael Reedy, Knut Schmidt-Nielsen, and Kathleen Smith have provided advice on lots of specific points. Jane not only kept her humor and perspective during the writing, she materially aided and abetted in her professional capacity as a reference librarian. Word processing was once again supervised by Coriander Cardamom Cat, who on one occasion when the screen read "press any key," immediately did so. And I am especially grateful to the Duke University Preventative Approach to Cardiology ("DUPAC") and Center for Living, purveyors of inspiration, perspiration, corporeal attenuation, and occasional jollification.

Durham, N.C.   S.V.
May 1991

Contents

VITAL CIRCUITS

1 *Plumbing Ourselves*

Some things are meant to be unobtrusive. We judge our electrical outlets, water heaters, and garbage collections satisfactory when they demand neither attention nor action. And so it is with our inner machinery in general, our circulatory systems in particular. What could be more distracting than worrying, moment to moment, about the proper beating of our hearts and the routing of our blood? Fortunately we can quite reasonably ignore such matters and not risk accusation of insufficient moral fervor by people with a more righteous level of concern—we're not talking about pollution or child abuse. There's no doubt that the proper functioning of our pipes and pumps does have an immediate urgency well beyond that of almost any of our other bits and pieces. Nonetheless in the well-regulated body such matters look after themselves without fuss and bother.

Things do go wrong, but having said that, I'll not trifle further with your anxieties by offering some distressing account of mishaps and disfunctions. Most of the time things go right—magnificently right—and so the present emphasis is emphatically positive. The point I mean to make is that, quite beyond being necessary during every minute of our lives, circulatory systems are among our greatest corporeal glories. I mean to talk about that splendor of design, its overall coherence, its exquisite detail, and its unobtrusive operation. I mean to celebrate the proper operation of a circulation—what it does, how it does it, how its arrangements have been adjusted to the varied requirements of different animals, how it retunes its operation as our own demands on it change.

As well as being about the magnificence of one's personal plumbing, this book is about science and scientists. Circulatory systems will be pressed into service to illustrate the kinds of questions we ask, the way we go about building a picture with the answers we obtain, and, most importantly, the very human and ordinary character of the enterprise. They'll carry, as well, the still greater task of showing why we find doing science so endlessly fascinating, and how that fascination transcends any question of practical application.

It's a book written by a biologist, and the outlook or bias of a biologist ought to be made explicit at the start. It is that circulation makes most sense and is most fun to talk about in a biological rather than in a more limited anthropocentric or medical context. While it's mostly about humans, we'll play the part of animals, and other animals will wander in and out. As even the most ancient investigators appreciated, the occurrence of common features amid the diversity of animal species can be of enormous assistance in figuring out how we work. As more recent generations have realized, a focus on animals that manage under severe or unusual circumstances can often reveal mechanisms too subtle for investigation in more everyday creatures. There's rationality in our practice of teaching aspiring physicians biology before we introduce them to medicine.

This particular biologist makes two claims to the inside story, one scientific and the other personal. It turns out that the kind of biology I do is intimately bound up with fluid mechanics. I worry about how animals and plants respond to the ways air and water move. Over the past few decades I've looked at such matters as how the shape of leaves reflects problems of keeping cool in low winds and avoiding damage in high winds, how a prairie dog can use local wind to ventilate its burrow, and how a swimming squid can use flow past itself to help refill for its next jetting squirt. And circulatory systems are at heart fluid mechanical systems of pipes and pumps. At least that's the additional bias of a biologist prone to physics-envy, as opposed to one who draws more heavily on, say, chemistry or geology.

On a more corporeal level, my own circulatory system has not always worked with quiet, unobtrusive efficiency—in recent years it has run my life in an awkwardly immediate and blatant manner. When someone with my professional predilections has to pay attention to one's own system, the combination leads to thinking about the mechanics of circulation as well as about how to stay in circulation. Three times each week, around dawn, I turn up at a cardiac rehabilitation clinic. There, as directed, I engage in activities that, quite without romance, make the heart beat much faster. As a biologist doing fluid mechanics, though, my outlook on these activities is a little different from those of either my fellow patients or the medical types who keep track of us. Thus (and speaking of tracks), I have now run around a certain track 7000 or 8000 times, something I find both boring and uncomfortable, but that I believe makes eminent biological sense. That provides motivation and comfort, offsetting my visceral suspicion that anyone who claims to enjoy running might do with a bit of psychiatric help. I hope some of that biological sense emerges from these pages.

My special concern, however, is no more than a mild exaggeration of something of a species psychosis concerning circulatory systems in general and hearts in particular. After all, the heart is our main metaphor for life itself. Hearty, heartfelt, heartrending; faint of heart, a noble heart, pure at heart—it's easy to go on, but I prefer to play social scientist and do a survey. For an unbiased set of respondents I have only to reach for my copy of Bartlett's book of familiar quotations. Sure enough, no organ approaches the heart's frequency of citation. I find only one quotation mentioning (favorably, in fact) kidneys, two for liver, four notes of lungs, twenty-four allusions to stomach, thirty-one uses of brain, all of sixty-one for breasts (crassly lumping the sexes), but no fewer than 581 quotations with some version of the word *heart*. If anyone is still insufficiently sanguine, there are a further 125 mentions of blood and fourteen of veins.

This association of heart and circulation with life has never been more apt. As a cause of death, infectious disease is now demographically minor in well-fed, decently housed, and expensively medicated first-world countries. What finally does in more of us than anything else are the diverse failings and functional afflictions of our cardiovascular systems—twice as many of us, in fact, as succumb to the various and more feared cancers. Still, let's put aside such depressing details—I meant it when I promised to talk mainly about how we work and not to pander to our fears of failure by dwelling on the occasions when we're in need of repair. What you'll get here is intended as perspective, a framework in which to put the information encountered elsewhere. Our plumbing should be no leaden subject.

* * *

Without further preambulations, let me introduce the present protagonists and their assigned roles.

The Task of a Circulation

Imagine a vast collection of ordinary suburban houses—quite a vast collection since we intend them as analogs of our 100,000,000,000,000 (100 trillion)[1] cells. Each has a certain functional autonomy, being equipped with the full facilities for storing, preparing, and eating food; for dressing and sleeping; for (as no biologist can forget) conceiving; and so forth. The autonomy, however, has its limits, transcended by the need for delivery of water, food, and energy, and for disposal of refuse and sewage. The biological analogs of these inputs and outputs are what a circulatory system manages—clearly crucial and just as clearly the most pedestrian of matters. Again I emphasize that for neither the houses nor the cells are they dramatic matters of active concern. In both cases requiring attention is the sort of pathological situation one prefers to avoid. For either, functions such as communication are more obtrusive—for a house both the wired and wireless sorts of communication, for a cell both neural and endocrine signals. Signals most often indicate changes, changes commonly demand responses, and we've learned (in several senses) to pay attention. Nonetheless, while perhaps more obvious and insistent, these communicative systems are only intermittently critical; their failure is less certainly and immediately fatal.

A circulatory system, then, moves material around within an animal. A carrier, blood, goes around and around, while other items make less monotonously repetitive circuits. Just what items are moved around varies a bit, both from time to time and from organism to organism. The system turns out to be more multifunctional than at first it appears. For one thing, the contrast I made with communicative systems involves no single, sharp distinction—hormones are commonly blood-borne, and they are nothing more than messages that regulate and synchronize the activities of otherwise fairly autonomous cells. For another, heat is transferred through movement of blood—it's a byproduct of all of our metabolic chemistry and is eventually dumped off upon our surroundings. While heat is produced throughout a body, it must be lost across a surface. Thus it needs transporting from core to periphery; or, at least under some circumstances, one's core will get intolerably warm. Under other conditions we find it useful to heat our appendages with what we make in the middle: cold fingers and toes are as likely to reflect poor circulation as to indicate insufficient heat production. Circulatory systems are even used to transmit force. The obvious case is erectile tissue: without heart and bloodstream we'd have no way to pump up our kind of penis—muscles have no role in the job. A less obvious but more widespread use of the hydraulic force of a pressurized bloodstream is in pushing blood though the filtration stage of kidneys.

By far the most demanding circulatory task for creatures like us is the transport of dissolved gases. In particular, oxygen is carried from the lungs elsewhere, and carbon dioxide, the main product of using that oxygen, is returned to the lungs. Of the two, oxygen transport is probably the trickier, simply because less oxygen normally dissolves in watery liquids. (Oxygenated rather than carbonated beer would be a lot less bubbly.) All the numerous tasks beside transporting oxygen and carbon dioxide could be done with a simpler and more sluggish system. I defend this assertion that gas transport is the most demanding function with a typical bit of biologist's logic—through comparison with a circulatory system that proves completely adequate for organisms that don't use it to transport dissolved gases. As it happens, insects have a separate system of pipes that carry oxygen and carbon dioxide to and from individual cells in (mainly) gaseous form. They do have hearts and some blood vessels, but their circulatory systems work at rather low pressures and flow speeds. To anticipate any argument that insects are primitive or inactive let me point out that many of them fly, and flying is about the most energy-intensive activity known either in nature or in human-designed transportation systems.

Among animals, we birds and mammals are unusual (but not quite unique) in maintaining rather warm and fairly constant internal temperatures. By insisting on having hot innards, we're committed to a rapid pace of life. We eat copiously and often,[2] and we release heat by combining stored fat and carbohydrate with oxygen. Our internal fires burn intensely, and not just as analogy or metaphor. When the fat's in the fire it uses the same amount of oxygen and releases the same amount of energy as when it's burned under much more controlled conditions within our cells. A sedentary person consumes about 70 kilocalories[3] an hour, which translates into a heat production of 80 watts. By contrast, a sedentary alligator of the same weight but with a body at ambient temperature consumes only about 17 kilocalories an hour, corresponding to 20 watts, about four times less. Our cells need proportionately more oxygen and produce more carbon dioxide than do those of the alligator; our circulatory systems have to bring the first to the cells from our lungs and take the second back to the lungs. Still, as we'll see, without circulatory systems even alligators are quite out of the question.

Alligators and insects are certainly not everyone's favorite creatures. But notice the heuristic utility of the perspective they permit, even if one's concern is entirely with humans. I've used the comparison with insects to argue that gas transport is the most demanding circulatory function and with alligators to suggest that the demand is especially great because we're warm-blooded animals. Thus I justify what might otherwise seem a disproportionate emphasis on a single function in a peculiar subset of vertebrates in all that will follow.

Simple Circuitry

At this point, we should take a look at a specific circulatory system, perhaps starting with a picture of our own, but a formidable problem immediately arises—we can't draw a decently informative illustration of our circulatory system on an ordinary printed page. For one thing, the system is three-dimensional; worse yet, it's made of parts awkwardly diverse in size. If capillaries, 8 micrometers (a three-thousandth of an inch) in diameter, are shown, then the heart (less than 5 inches high and slightly more than 3 inches wide) won't fit on the page. So it takes considerable imagination to visualize the thing. In fact, there's a more general problem. We biologists are all too prone, I think, to submerge ourselves and everyone else in anatomical and terminological detail. It's better if we defer that plunge at least briefly. Beginning with an overall view of function will greatly reduce the burden imposed by a lot of names and places.

For now, describing function can be helped considerably by using analogies and by beginning with less complex biological cases. So we'll go back to our suburban houses, now using them to draw parallels between household heating systems and the oxygen distribution systems of animals. In particular, we'll focus on those systems in which heat is produced centrally and then distributed as hot air or hot water to the various rooms. The components of such a system are, first, a set of pipes for transporting the hot air or water. There must also be some pump for either circulating medium. Finally, two different kinds of exchangers are needed. In one, heat is acquired by the medium—we call it a furnace; in the other heat is transferred to the living quarters—it may be a set of radiators or just a trivial mixing arrangement for hot air coming out of some ports. So—pipes, pump, and two kinds of exchangers. These are precisely, and for just the same reasons, the indispensable components of circulatory systems. All others are really just fillips and flourishes. At this point we might draw a picture of a heating system on one side of a page and at least a crude one of a circulatory system on the other, but the very different appearance and location of the parts of each would get in the way of recognizing any underlying similarity in their interconnections. A better way is to borrow the engineering practice of using purely functional diagrams. The hot air heating system (Figure 1.1a) needs a full set of distributional ducts, but it manages to work without much in the way of return pipes. A house has a fixed volume, and air is pretty nearly incompressible at any speeds of flow with which we might comfortably coexist. So blow air from furnace to periphery, and the air will return to the furnace of its own accord. A little help may be needed—my house has a few return ducts from the more remote bedrooms, and it ought to have another to keep the living room (on the other end) properly heated. (And animals arranged this way come equipped with a quite a few odd ducts and auxiliary pumps to handle their various anatomical tortuosities.)

The circulatory system of a snail or spider (Figure 1.1b) is arranged

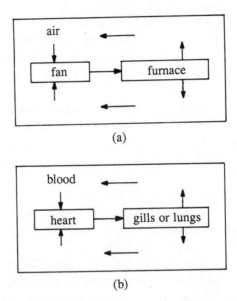

Figure 1.1. Analogous arrangements—(a) a hot air heating system, and (b) the "open" circulatory system of a spider or snail.

in much the same way as the hot air heating system. Blood is pumped by a heart; it exchanges dissolved gases with the environment through a set of gills (most snails) or sheet-like lungs (most spiders); and it moves heartward again without benefit of very specific piping. In this heartward oozing through the rest of the body, the second exchange takes place, the one by which tissues are provided with dissolved oxygen and dump their dissolved carbon dioxide. In the biology textbooks of my youth such a system, termed an *open* circulatory system, was derided as primitive on account of its lack of discrete return pipes. That's probably an unnecessary and inappropriate bit of vertebrate or mammalian hubris. Spiders, lobsters, and the rest of the great phylum of arthropods (which includes the insects, a distinct group from the spiders) have inextensible outer shells rather than our more flexible skin. Furthermore, water and all bloods are even less easily compressed than air. If blood is pumped to the front, it simply has to find its way rearward since it can go nowhere else. Discrete pipes to carry blood rearward would be just so much needless baggage, wasting space better invested in equipment closer to proper reproductive imperatives—in short, to guts and gonads. Similarly, pumping blood outward to the tips of legs must, if no leaks develop, force it back inward automatically.

By contrast, in a recirculating hot water system (a fine, if initially expensive way to heat a house), a pump pushes water first through the heat exchanger of the furnace and then through the heat exchangers more commonly referred to as radiators (Figure 1.2a). A return pipe from each radiator carries the somewhat cooler water back to the pump and furnace. Radiators are necessary, of course, because our houses are filled with air rather than water. The circulatory system of a fish (Figure 1.2b) is closely analogous. A heart pumps blood forward; blood then passes through the gills where it acquires oxygen and disposes of carbon dioxide. After that it goes by way of arteries to the other exchanger, the network of capillaries elsewhere in the body; it finally returns through veins to the heart again.

The Bird-and-Mammal Scheme

What of ourselves, neither spiders nor fish, but proper mammals? We use a modification of the piscine scheme; and I mean modification in both of the biologist's senses—historical and functional. That is, our structures represent the end of a long history of modification of what some early fishes invented about 400 million years ago. Our mammalian version probably reached very much its present form by a still ancient 200 million years ago. At least that's the well-accepted inference, since our circulatory equipment operates in a way quite similar to that of modern fishes and most likely of ancient fishes as well. A little educated guessing is, of course, involved—circulatory systems don't leave nice, accommodating fossils as do bones. It would, however, be quite astonishing if ancient fish did things very much differently from their extant descendants. The evolutionary history isn't irrelevant. If you set out to design, de novo, a mammalian circulation, you might do better to arrange it a bit differently from what we in fact have. If you're limited to modifying pre-existing anatomy, then you're likely to end up with certain quirks and jury-rigged arrangements.

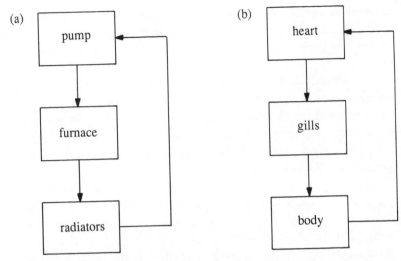

Figure 1.2. Another analogous pair—(a) the layout of a hot-water heating system, and (b) the arrangement of the circulatory system of a fish.

It's worth noting that, while descent with modification is evolution's way, existing structures are not uncommonly pressed into service for new functions. The main buoyancy devices of fish, swim bladders, come from the same antecedent structures as do our lungs. The tiny bones in our middle ears come from the same antecedents as some associated with the gills of fish, bones whose equivalents see service in the jaws of reptiles and amphibians.[4] By contrast, these circulatory systems we're talking about not only come from the same ancestral structures, but also retain the same essential function. Vertebrates have been pretty conservative, evolutionarily, and circulatory systems seem to have especially strong traditions. According to the best evidence, birds and mammals represent separate offshoots of the reptilian stock, lineages that branched from different reptilian forebears at different times. Birds and mammals nonetheless hit upon the same basic alterations of the reptilian circulation. Perhaps that isn't too surprising—the changes are simple, logical, and effective.

Figure 1.3 is a circuit diagram of the circulatory arrangements of mammals and birds. Two really major alterations are evident. Gills, suitable for extracting oxygen from water, have been replaced, func-

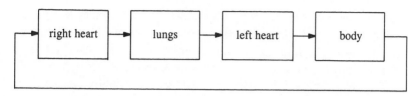

Figure 1.3. The path of the blood in birds and mammals.

tionally, by lungs that extract oxygen from air. Gills haven't, it turns out, been modified into lungs. Curiously, the components of the gills have been lost or put to other purposes, and lungs have developed separately as outpocketings of the esophagus—the outpocketings that in fish we noted serve as gas-filled swim bladders. The other alteration is the addition of a second heart. A fish pushes blood through two exchangers with one heart; a bird or mammal has an additional pump between the exchangers. Our blood goes from one heart to the lungs to the second heart to the rest of the body and then back to the first heart.

At once something distinctly queer about the arrangement jumps out. Two hearts located in quite different places hardly constitute a proper vision for St. Valentine, and they certainly don't correspond to our common notion of what we have inside. No mistake—in a functional sense each of us does have two hearts, but they just happen to be combined into one muscular organ, and they can beat only in synchrony. The anatomical combination obscures the functional distinction. That combination, though, is probably one of those accidents of "descent with modification," as Mr. Darwin put it, one of those odd quirks evolution leaves in its wake. Anyhow, half-heart one (right side) pushes blood through the lungs, and it's a little smaller; half-heart two (left side) is larger and pushes blood through the rest of the body.

The scheme makes good sense. As we'll see later, these pumps really expend energy pushing blood through the tiniest vessels, mainly capillaries, of both the lungs and all the other organs. In a plumbing system, loss of energy is associated with a decrease in pressure, which will be discussed further in Chapter 4. Therefore the pressure imparted by the pump to the blood decreases substantially as blood passes through a set of tiny vessels. Going through two sets in sequence (at a given speed) would require that the blood start with twice the pressure needed to push through one set. Among other difficulties, the pipes leaving the pump would need additional reinforcement or would be more susceptible to blow-outs (ruptured aneurysms, in the trade). Solution—a booster pump between first and second sets of tiny vessels.

(Again, to dismiss the single-hearted fishes as primitive would be an indefensibly anthropocentric attitude. Fish do quite nicely, thank you— they're the most numerous of all vertebrates. They, like alligators, just don't ordinarily go in for our profligate rates of oxygen use. Fish blood circulates more slowly than does our own, and the pressures involved are correspondingly lower.)

Which pump we call the main one and which the booster is more or less arbitrary. Our "lung" half-heart (right side) is the more immediate functional descendent of the sole piscine pump, so we might best regard the larger "body" half-heart (left side) as the secondary booster. Curiously, the creatures that stand between fish and us in an evolutionary sense, amphibians and reptiles, have various sorts of in-between arrangements about which more will be said later. The full, deluxe, duplex deal graces only birds and mammals. Odder still, these nearly identical hearts clearly represent two quite independently evolved modifications of the hearts of reptilian forebears. Possession of these fancy double hearts is surely related to the fact that only birds and mammals routinely maintain body temperatures well above those of their surroundings.

A few items of terminology are now unavoidable. (I only disparage terms as jargon when I don't need them.) Our mammalian system consists of two pumps, collectively called the heart, and three main kinds of pipes. *Arteries* have thick walls that withstand the full pressure generated by the squeeze of the heart; they carry blood away from the heart. (The largest artery, the one leading out to the body from the heart is called the *aorta*.) *Capillaries* are tiny pipes with very thin walls that are supplied with blood by the arteries. Whether in the lungs or elsewhere, they're the sites at which material is exchanged between blood and the tissues of the body. *Veins* have thinner walls and much lower internal pressures than the arteries. They drain the capillaries and return blood to the heart. Each side of the heart consists of two pumping chambers serially arranged, an *atrium* (sometimes called an "auricle" on account of being a bit ear-shaped) and a *ventricle*. Thus the heart has four chambers in all. The system has other components, but these are the main ones.

What flows through the system is blood, consisting of a watery liquid (*plasma*) in which are suspended a variety of small entities—mainly cells of varying sorts and sizes. In addition, blood carries dissolved gases, salts, hormones, droplets of fat, energy-yielding molecules such as glucose, waste materials such as urea, and a wide variety of proteins involved in such functions as clotting and inflammatory responses. These materials enter and leave the blood as it passes through the capillaries of the various organs.

Quantifying Things

Let's add some numbers to all the words, using the human circulation as source of data—adult humans of average size. At least in pushing blood around humans are not especially unusual mammals, lots of human data are available, and both reader and writer are without the slightest doubt quite human. While most of these numbers will come up again later, it will be handy to have them in one place.

Blood volume: 5.2 liters (5.5 quarts) in the body, about 5 percent of which is in the heart. Being relieved of a pint (half a quart or about half a liter) is no big deal if you're in normal health—it's less than 10 percent of your stock. Lady Macbeth, you may remember, found this large volume noteworthy.

Output, left ventricle: 5 liters (5.3 quarts) per minute for a person at rest, and about six times higher in strenuous exercise. The output of the right ventricle is, of course, precisely the same. When at rest, the whole blood volume of the body passes through the left heart every minute, or a bit of blood takes about a minute to make the complete circulatory circuit. During heavy exercise, the circuit takes only about 10 seconds.

Stroke volume and heartbeat rate: The 5 liters per minute is the product of about 70 milliliters (2.5 ounces) per stroke times about 72 strokes per minute.

Fraction of blood occupied by cells ("hematocrit"): from 37 to 52 percent. Most of this volume is occupied by the so-called red blood cells, of which you have 5 million per cubic millimeter or about 25,000,000,000,000 (25 trillion) in all.

Speed of flow leaving heart: 0.3 meters (about 1 foot) per second, resting. In maximal aerobic exercise in a well-conditioned person it rises to about 2 meters per second (about 6 feet per second or 2.5 miles per hour); a leisurely walking pace.

Speed of flow in capillaries: 0.4 millimeters per second at rest or about 800 times lower than the speed at the aorta, the exit from the heart. That's 50 days per mile, in our perceptual world exceedingly slow.

Overall area of capillary wall: 8000 square meters or all of 2 acres. That's the surface across which material can move between the circulatory system and the surrounding tissues.

Combined length of pipes: 100,000 kilometers (60,000 miles)—more than twice around the earth at the equator.

Heart weight: About 0.3 kilograms (11 ounces) or about half a percent of body weight. A heart isn't an especially large organ.

Power output of heart: 1.3 watts at rest, about 8 watts during exercise. The latter is about 30 times *less* than the power per unit weight of a good internal combustion piston engine of almost any size.

Power consumption of heart: At rest, this is 13 watts, or about a sixth of the body's resting power consumption of 80 watts, as mentioned earlier. 1.3 (earlier) divided by 13 is, of course, 10 percent. Thus about 10 percent of the fuel (fat, etc.) consumed by the heart appears as useful, blood propelling, output.

And Questions Left Hanging

I've now sketched the main features of the circulatory systems that will occupy the forthcoming pages. All the rest, though, isn't just detail. The present description is about as bald as its author, only a little more user-friendly than a textbook, and without all of my favorite devices. Worst of all, it conveys little sense of the logic and elegance of the various features of these systems. To push one of my long-standing polemical themes, such features do not unfold from an evermore detailed study of the organisms, but from consideration of just what problems these circulatory systems are up against. It sounds backward—to imagine problems and then to investigate what animals have done about them—but the approach turns out to be useful both as a way to do science and as a way to talk about functional systems. The following, then, are a few of the problems.

- Liquids are, for all practical purposes, incompressible. If, in contracting, the heart's chambers get smaller, then blood pushed out must turn up as an increased volume elsewhere. One attractive fix, beating left and right hearts alternately, isn't used, perhaps because of the fact that the two ventricles are squeezed by the same muscle (—two hearts beat as one, even without romantic hyperbole). Are there functional consequences of this periodic variation of the volume of the rest of the system?

- If you blow into a cylindrical balloon it inevitably expands almost to full inflation in one portion before inflating anywhere else. The initial bulge is what we'd call an *aneurysm*. When you have one in any part of your circulatory system you're in mortal danger since aneurysms, like balloons, are prone to burst. Blood vessels don't ordinarily develop aneurysms, however, and one wonders why. They certainly have stretchy walls—the situation described in the previous item absolutely demands stretchy walls.

- Your circulatory system is a serial arrangement of pipes and pumps (as just described). Thus in any period of time as much blood must pass through the lungs as passes through the capillaries of every other organ in the body combined. Lungs are only a tiny fraction (about 1 percent) of the body's mass—are they really as bloody as this requirement seems to imply?

- A heart is mainly muscle; contraction of this muscle reduces the volume of the chambers inside. So blood gets squeezed out. Valves ensure that it always get squeezed out in one direction, and each valve dutifully opens and shuts once in each stroke. How can they perform their mechanical task when they have no nerve supply to ensure that they open and shut at the right points in the stroke cycle?

- Blood flows rapidly in the arteries but much more slowly in the capillaries, allowing time for diffusion of dissolved materials across the capillary walls. But the blood speeds up again as it gets into the veins and is finally heartward bound. Where's the pump that so speeds venous blood on its return journey?

- The hearts of large animals beat less frequently than do those of small animals. Large animals live longer, usually, than do small ones. If you multiply the number of heartbeats per minute by the number of minutes per lifetime you get a figure for heartbeats per lifetime, certainly an odd datum. What makes the datum really strange is that its value, about 1 billion, is nearly the same for most mammals. Anything constant amid the great diversity of organisms catches our attention—but what functional significance, if any, might attach to *heartbeats per lifetime?*

It's easy to go on, but you surely get the idea. Most of a book lies ahead, and with some heartwrenching cardiac pun (your choice) we press on.

Notes

1. I'll use the U.S., not the U.K., naming conventions for large numbers throughout.

2. A cat dealing with even a badly mouse-ridden house still has to be fed. By contrast, it takes an awkwardly large number of geckoes to keep the cockroaches at bay in a tropical home.

3. The *calorie* (or *Calorie*) used in the nutrition or diet business is technically a *kilocalorie*, 1000 of the calories of the physicist. I'll use kilocalorie to avoid any ambiguity. In fact, the preferred unit for energy in contemporary scientific practice is neither, but instead something called the *joule* (after James Joule, 1818–1889), equal to about a quarter of a calorie or a four-thousandth of a kilocalorie or one watt-second. A joule is thus the energy used by a one watt bulb kept on for one second.

4. As our bio-poet, John Burns, described the evolution of auditory ossicles, "With malleus Aforethought Mammals Got an earful Of their ancestors' Jaw."

2 *Pumps and Pipes*

In our unguarded moments, we biologists may admit a certain ambivalence about structure. On one hand it is the classical guts of our subject. At some level we carry the conviction that the fantastic structural complexity of cells, of organs, and of organisms wouldn't have evolved without some functional concomitants driving the process. After all, making some functionless structure is not going to improve a creature's procreative potential, and the latter is the very crux of natural selection and evolutionary change. On the other hand, that same complexity of living structure proves in practice to be a great curse for those of us trying to see common functional features amid the wild diversity of living systems.

Biology made its place as a respectable science by naming and classifying creatures and by describing structure; only secondarily (and less respectably) did it speculate about function. (About biology, for instance, Aristotle's explanations were right so infrequently that one suspects pure coincidence.) My own conviction is that this vaguely historical sequence, moving from studying structure to deciphering function has no special virtue for efficient and engaging explanations. It seems better to begin with a broad view of functional systems, using diagrams, models, and analogies—and only thereafter to get into anatomical detail and terminology.

Further Functional Notions

We've seen that a circulatory system of our sort is made up of a bunch of pumps and pipes. By far the most important pumps are, of course, the two comprising the heart—they're the active components in the sense that only the heart requires a substantial supply of energy to operate. Still, as we'll see later, the heart is by no means the only pump in the system. The pipes are those arteries, capillaries, and veins mentioned in the last chapter—they're essentially passive components, "mere" conduits. While many of these vessels have muscle in their walls, the muscle isn't used to power a pump. Confusing—hearts are muscular pumps but aren't the only pumps; vessels may be muscular, but they don't pump. This was historically confusing, too. Arteries pulsate in opposite phase to the ventricles: when the ventricles contract, the arteries swell and you can feel a pulse wherever an artery runs just beneath the skin. That changes in arterial size are just passive consequences of the heartbeat was not, and is not, self-evident. True, yes; obvious, no.

Let's consider hearts first as simple squeezers and then look at their peculiarities as pumps. A heart is basically a bag with a muscular wall, and a muscle is an engine that puts out power by actively shortening. Upon receipt of an appropriate signal, it tries to pull its ends together—it generates a tensile force. If one or both ends are free to move, the muscle gets shorter—it contracts. If the muscle is wrapped around a squeezable container, then contraction reduces the diameter of the container. The engines of our technology work by either expansion or rotation, so one can't easily build a mechanical heart that works in the same way as a real one.

Putting the squeeze on a chamber using muscle has an oddly practical aspect that often escapes notice. Muscles shorten forcefully, but not especially far—a muscle ordinarily shortens only to something still well over half its resting length. Does this mean that in a beat a heart can at best squeeze out less than half the blood it contains? A model based on simple solid geometry suggests an answer. Imagine a sphere that can make its walls shorten. If, say, the wall circumference were to decrease by 20 percent, that would decrease the radius of the sphere by 20 percent. But the volume of a sphere doesn't follow its radius in such simple fashion—you may recall that volume changes with the radius cubed,[1] that is with radius twice multiplied by itself. The upshot is that a reduction in circumference of 20 percent gives a reduction in volume of fully 50 percent. Figure 2.1a represents the matter in a two-dimensional diagram.

Very roughly, then, if such a heart were to contract its muscle by 20 percent it should pump out half the blood it contains. That's about what happens during the normal heartbeat of a person at rest—2.5 ounces pumped, with the same amount remaining. In strenuous exercise the heart muscle contracts further, and it manages to squeeze out fully 78 percent of the blood it contains. If it worked just like our crude spherical model the heart would have to contract its muscle by only 40 percent of its resting length to do so.

Still, even 40% is awkward. A muscle develops its best power with much less drastic shortening, so our model asks the heart to run rather inefficiently just when called upon for its best effort. Adding two additional elements of realism to the model turns out to help matters. Assume that the squeezing sphere has thick walls and that these are incompressible, both quite reasonable for hearts. Assume further that at rest the volume of wall is twice the volume of the chamber inside, as shown in Figure 2.1b. The calculations are only a little more complicated; they reveal that to expel half of the contents of the chamber, the outer layer of muscle need shorten only by 6 percent. To expel the entire volume of the chamber, the contraction needs to be only about 13 percent. Both of these are very nice figures from what we know of the performance of muscle—they ask that heart muscle operate in a range of shortening in which muscles work well.

I go through this slightly artificial exercise to make a general point as well. Models get used all through science. In essence, a *model* is a hypothesis about how some real thing works. To the extent that the model's performance matches reality, we have evidence (not proof!) that the model is a good one. Its behavior might even be used to predict unknown aspects of the behavior of the real thing. Often a lot of insight can be obtained by considering a series of models running from especially simple and abstract to fairly realistic but more complex. Much can be inferred from the places in the series where particular problems first emerge or where performance takes on specific aspects of reality.

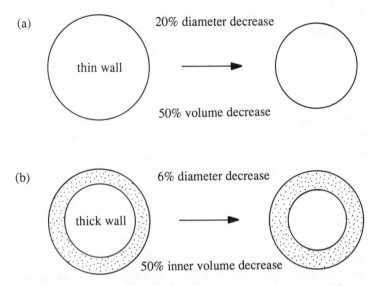

Figure 2.1. (a) A small change (20 percent) in the diameter or circumference of a spherical shell gives a large change (50 percent) in the volume enclosed. (b) If we assume a massive, incompressible wall, an even smaller change (6 percent) in circumference gives the same change in the volume of the central chamber.

Creating the series is a little like planning a construction project, with a sequence from sketch to scale drawing to specification of procedural details; the main difference is that our series of models has an analytic rather than a synthetic goal.

To go much further, it's handiest to put aside simple, squeezing spheres and turn to hearts themselves. Earlier, the existence of entrance chambers attached to a heart was mentioned without comment. These atria are invariably found in the hearts of our fellow vertebrates, and they're parts of the hearts of many of the active kinds of other biggish animals. They're most often muscular chambers, but they are never anywhere near as massively and powerfully muscular as are ventricles. Despite this difference they pump blood at precisely the same rates as the ventricles to which they're attached.

What seems to be going on is more or less the following. A pressure difference is what drives flow through circulatory systems, or through any other plumbing for that matter. More explicitly, an initial high pressure provides the push, and a heart provides that high pressure. A pressure difference, though, demands that pressure at the other end be low—again, it's the difference that matters. High pressures, as we'll see, are not without costs and dangers; and these argue that the pressure of blood returning to the heart should be kept as low as possible. Yet another problem thereby arises. How can the main pump, a ventricle, be rapidly refilled if blood returns at low pressure? It cannot actively expand, as we'll see shortly, so blood must be forced in. And that's where these weak little auxiliary pumps, the atria, come in. They are, in a word, *superchargers*. Figure 2.2 shows the arrangement schematically.

For people who fondle (at least mentally) automobile engines the idea will be nothing new. An ordinary car sucks its explosive mixture into a cylinder because the piston goes downward and reduces the local pressure below ambient. So the fuel and air[2] enter at this low pressure, and the useful pressure caused by combustion builds from this low ebb. One can extract more power if an external pump, a supercharger, is used to force in the mixture. The mixture is compressed, so more goes in, and the baseline pressure before combustion is higher. Vertebrates and mollusks—two of the three great culminations of complex animals—have separately evolved atrial superchargers. The main difference between hearts and fancy cars is that hearts use the supercharging to increase the volume per stroke more than the maximum pressure. More mixture enters because ventricles are stretchy not because the fluid is compressible (blood isn't).

Back to muscular pumps. A muscle can only contract; if it surrounds a chamber its contraction can increase the internal pressure in the chamber, squeeze out some of the contents, or both. A little more machinery is needed to make a practical pump that does more than merely stir things up. In practice, nature uses two versions of that "little more machinery" in making muscles run pumps. (Some living pumps have other power sources, but we'll not be distracted by them here.)

One arrangement is the one used by our intestines. It involves waves of contraction running down the length of the pipe, each wave pushing a bolus of fluid ahead of it. The scheme is called *peristalsis* (Figure 2.3), and it approximates what you do when trying to get the last bit of toothpaste from the bottom of the tube. The same machinery can push

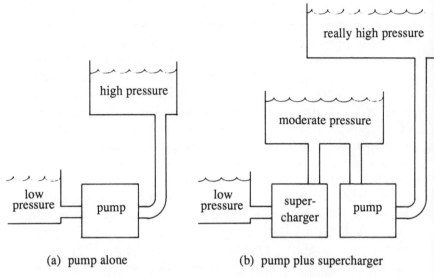

(a) pump alone   (b) pump plus supercharger

Figure 2.2. A hypothetical pump before and after the addition of a supercharger or atrium: (a) pump alone; (b) pump plus supercharger.

fluid in either direction, and it can deal with sludgy fluids, but it's relatively inefficient in its use of power. Peristaltic hearts exist, but in neither fish nor fowl nor, for that matter, in any other vertebrate. Some worms have them, as do a curious group of creatures in the same larger lineage as vertebrates—they're called sea squirts, or sea pork, or tunicates, or ascidians—that for unknown reasons have hearts that periodically reverse their direction of pumping.

What our hearts do is quite different. Those supercharging auxiliary chambers, the atria, squeeze and drive blood into the ventricles. The ventricles then give discrete, simultaneous squeezes, and blood flows out into the arteries. The two sounds you hear when you put your ear against someone's chest indirectly reflect those contractions—the gentler atrial contraction followed by the stronger ventricular contraction. But what prevents ventricular contraction from driving blood back into the atria and the veins as well as into the arteries? And what prevents the ventricles from being recharged from the largest arteries rather than from the atria? In short, what makes the pump decently directional?

Consider the squeeze-bulb of a kitchen baster as a crude model of a

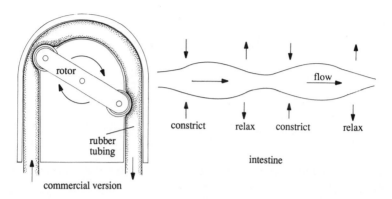

Figure 2.3. Pumping by using traveling waves of constriction in machine and intestine. Neither system is especially efficient in terms of energy, but the scheme has other compensating virtues. Peristaltic pumps will push semi-solid slurries, and by staying in a pipe the contents avoid contamination in valves and chambers.

muscular chamber such as a ventricle. What is the minimum additional machinery necessary to make it into a one-way pump? One answer, and I think the simplest, is a pair of "check valves"—devices that allow fluid to pass in only one direction. Arrange the squeeze-bulb and check valves as in Figure 2.4 and you have a one-way pump. A single valve is better than none, but it allows a lot of back flow; more than two accomplishes nothing further. One neat feature of this scheme may escape your notice since what's neat is that something proves unnecessary. An automobile engine has valves to let air and fuel into the cylinders and to let exhaust out of them. Timing the opening and closing of all those valves must be carefully controlled by timing belt, camshaft, lifters, and so forth. Both the baster with check valves and our ventricular chambers have automatically self-synchronizing valves! They open and shut in response to nothing fancier than the pressure changes caused by squeezing the bulb or contracting the chambers of the heart. That, incidentally, answers the question left hanging in the last chapter about how heart valves manage without nerves.

If one ventricle needs two valves, an input one and an output one, then two ventricles require four valves in all. And these we have. One might reasonably expect that the presence of the pair of atrial ante-

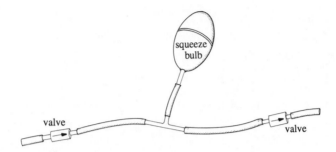

Figure 2.4. A simple pump made from a squeeze-bulb and a pair of check valves.

chambers would entail provision of two further valves. Reality is just a little more complicated. A valve is present between the vein from the lower part of the body and the right atrium but not between the vein from the upper part of the body and that same atrium. Nor are there ones between pulmonary (lung) veins and left atrium. Contraction of the atrial wall effectively closes these inlets, so functional valving is accomplished even without specific structures. Not only don't nerves supply our heart valves, the valves don't even need muscles. And if you have trouble with leaky or otherwise faulty valves, fairly good replace-ments can be installed. These replacement valves are either artificial, mechanical components, or else are valves "borrowed" from, most commonly, defunct pigs or people. Once installed, they're quite de-pendable. (One might say that they last a lifetime, but they sometimes do need replacement, and using the word *lifetime* in this context makes one a little uneasy.)

Looking at a Heart

At this point, I originally planned to advise the especially devoted or compulsive reader to abandon hearth and chair, to rush forth, and to purchase a heart from the nearest purveyor of such items. However, ordering an untrimmed heart from a butcher shop turns out to be not as simple as it used to be. Heart is the least fatty of any meat from the large mammals we commonly eat;[3] according to my informant at the nearest market, this explains its relative scarcity. Say a manufacturer of the sliced edible we euphemistically call "lunch meat" wants to make a product that can truthfully be labeled, say, "98 percent fat free" (e.g., 2 percent fat by weight or around 17 percent of total calories). Some source of very lean meat is almost mandatory, and heart is thus clearly desirable, especially since no toughness or unusual texture survives the grinder. By the same token, it's worth trying if you want to eat red meat but have to worry about your intake of saturated fat.

If you do want to take a look at the real item, be advised that getting an untrimmed heart is best—the atria and a bit of piping are left attached. A "trimmed" heart has only ventricles (the really meaty since the most muscular part), so that's the next best thing. Usually it's slit lengthwise in order, presumably, to lie flat between plastic wrap and styrofoam; but you can easily truss it up again. In any case whether you get a heart or a piece of one and are at all carnivorous, you really shouldn't waste it. But don't cook heart as if it were some commonplace skeletal muscle—I've provided a little more guidance and a recipe at the end of the chapter. In our household, beef heart is purchased, when available, as a special treat for the feline who supervises the establishment.

An alternative hands-on approach to a heart is to order an embalmed one. These are cheap and not as unpleasant as you might think. The biological supply companies do a good job at preservation, so the texture and even color are surprisingly lifelike, and they can be handled quite leisurely at room temperature. Also, the days of formaldehyde fumes are past—an embalmed heart has almost no odor. A few addresses are given at the end of the chapter. You usually have a choice of cow, sheep, or pig. No one is clearly preferable, inasmuch as beef hearts are nice and big, pig hearts are usually cheapest, and sheep (lamb, really) hearts are about the size of our own. Anatomical differences are trivial. Not that one absolutely must get quite so unvicariously immersed in the subject—illustrations do accompany the words that follow.

In a mammal, the volume of the chest or thorax beneath the ribs is mainly filled by a pair of lungs, one on each side and each in its separate cavity (Figure 2.5). Between them is a substantial wall, called the mediastinum (accent on the penultimate syllable), extending from ster-

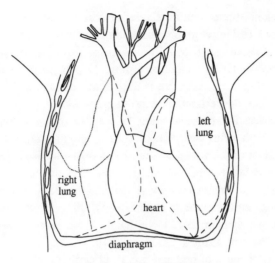

(a) the place of the heart in the thorax

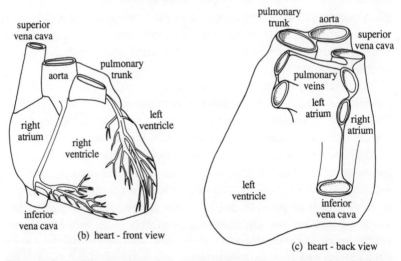

(b) heart - front view

(c) heart - back view

Figure 2.5. (a) The location of the human heart in the chest, with the items normally in front of it removed. Dashed lines mark the inner margins of the lungs. (b) The heart itself in front view, with its coronary arteries. (c) And the same from behind.

num in front to backbone behind[4] and from the top of the thorax to the diaphragm at the bottom. Down through that wall, in the back, runs the esophagus, carrying food from the head above to the stomach in the abdominal cavity beneath. And in that same wall the heart is located, a little to the left of the midline. It's located within the so-called pericardial cavity. Thus the heart's surface is not attached to the wall of the mediastinum, and it's free to slide around as the animal moves and the heart beats. The cavity contains a few ounces of fluid that, among other things, provides lubrication. When lubrication is inadequate a frictional rub—acute pericarditis, is audible through a stethoscope.

If you were to remove the sternum and a bit more of this and that, you'd see the heart from what we're calling the front, the view most often illustrated. (Pardon me if I'm getting a bit graphic—I do really understand that some people are normally squeamish.) This puts the ventricles down and the atria up, a little like a dangling strawberry or an upside-down pear. (If you have an intact heart, you ought to orient it as if viewed from the front—put the two biggest and sturdiest pipes upward and less well-defined connections around back.) Since you're facing the heart, its left side is on your right and its right side is on your left.

The right ventricle is mostly what you're now facing, with the left partially rotated around behind to the right. In fact the right ventricle wraps a bit around the left. As you can see from Figure 2.6, the right ventricle has much thinner walls, concomitant with the lower arterial pressures in the circuitry of the lungs. (The wrap-around arrangement and the very different thicknesses of the ventricular walls are quite obvious in a duck or goose heart.) The venous pipes leading into the atria are mostly hidden behind, and the pipes leading out from the ventricles emerge on top, between the atria. The left top pipe as we face it (on the right with respect to the body) is the great aorta, about an inch in internal diameter, which carries blood to the body from the left ventricle. The right top pipe (on the left of the body) is the pulmonary (lung) trunk, which branches into the left and right pulmonary arteries—one for each lung. The right atrium is visible at the upper left; the left atrium is mainly around behind.

The arrangement is almost certain to confuse you at first sight—

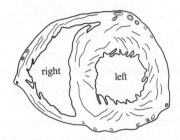

Figure 2.6. A cross section of the heart through the ventricles—to show the great difference in the thickness of their walls.

while the right atrium seems to be in the right place, it looks as if the right ventricle is on the wrong side. The problem is that the whole organ is twisted, obliquely mounted in the body, and mildly asymmetrical. Looking back at our earlier diagrammatic representation, Figure 1.3, on page 13, ought to keep you from getting too badly disoriented. I think anatomists (and many other biologists) have unusually well-developed visual perceptions that work well with oddly shaped, three-dimensional representations. The way I cling to diagrams (even when I taught anatomy) is probably an aspect of the same personal defect that almost drove me out of biology when a freshman course demanded proper drawings. Dissection was easy—my manual dexterity later proved adequate to attach little wires onto fruit-flies.[5] Nonetheless, I have horrible memories of my attempts to get images down on papers that would be graded.

So—blood goes forward into the atria, down diagonally into the ventricles, and upward as it heads out to lungs or elsewhere. Into the right atrium, thence to the right ventricle, and out to the lungs in the *pulmonary* circuit. In from the lungs to the left atrium, down into the left ventricle, and out the aorta to the rest of the body in the *systemic* circuit. Heart to heart, twice over.

A few more blood vessels, much smaller ones, are visible in the drawing of the front of the heart or on the surface of the real heart. These are the coronary arteries that supply oxygen and the other metabolic desiderata to the heart muscle. From our present vantage point, on the right lies the main branch of the left coronary artery. This branch

of the left artery conveniently marks the approximate boundary between right and left ventricles. On the left lies the right coronary artery, running in a groove between the right atrium and the right ventricle. Finally, we have some structures helpfully located! As you can
see from the drawing, both of these arteries receive blood directly from
the aorta. Adjacent and somewhat beneath each is the corresponding
vein, carrying blood back to a junction at the right atrium. The right
coronary artery branches near the bottom of the drawing, with one part
going along the bottom margin of the heart and the other going around
behind and out of view. The left artery branches near the top of the
heart, with the branch not shown going around behind. Thus the back
of the heart gets supplied by vessels coming in from opposite points
from the front—lower right and upper left instead of upper right and
lower left.

(Coronary arteries are prone to suffer internal narrowing as fatty
material gets deposited on their walls; the result is a reduced blood
supply to the heart muscle, not at all a good thing. The condition,
coronary artery disease, is extremely common, particularly in elderly, obese,
and inactive people who eat a lot of animal fat, and in those with strong
hereditary predispositions. Two invasive treatments are used these days.
One is *balloon angioplasty*, a kind of internal wall-smoothing. The other
is *coronary bypass surgery*, in which the narrowed vessels are bypassed
by new ones, spliced between the aorta and the lower reaches of the
coronary arteries. The new vessels are merely some surface veins from
the same person's legs, veins that are parallel to other veins and therefore mildly redundant.)

We can now turn the heart over, either figuratively (Figure 2.5,
again), or on your cutting board. Looking from the back, left on paper
is now really left and right is rightly right. In the middle is the left
atrium; below and off to the left is the left ventricle. The edges of right
atrium and ventricle are visible on the right and bottom. Details of the
pipes coming into the atria vary a little, even among such a conservative
bunch as the mammals, but the human pattern isn't in any significant
way deviant. (Humans may be special in lots of respects, but we should
be reminded that we have unexceptional mammalian circulatory systems.) Four large pulmonary veins come in from the lungs (two from
each) and separately enter the left atrium. And two even larger veins

carry blood to the right atrium, the inferior vena cava ("hollow vein") from the lower part of the body and the superior vena cava from the upper part (some mammals have a pair in place of a single superior one). That's it on the outside.

Inside (Figure 2.7), the most conspicuous item is, I suppose, the thick muscular wall that separates the ventricles. If you're working on a real heart, slitting it lengthwise will expose the interior of one or the other ventricle, the choice determined by just which side of the inter-ventricular branch of the left coronary artery receives your knife. The interiors of the two ventricles are similar; what is for some reason surprising is just how unsmooth are the inside walls. Folds and columns (*trabeculae*) of flexible but strong and inextensible tissue run across the ventricles, mostly in an up-and-down direction, as do some muscles (*papillary muscles*). Together, these assist in the operation of the heart valves and probably help prevent any overdistention of the ventricles.

The most interesting internal structures are certainly those valves. As explained earlier, they're critically necessary to get unidirectional flow, but they're nonmuscular, noninnervated, passively operating structures. While much has been made of differences among them in structure, to me they look quite similar. Two or three pieces of flexible sheet (*valvules*) extend inward from the circular periphery of an orifice, pieces large enough to completely occlude the opening (Figure 2.8a). If the pressure is higher on the normally upstream side, then the individual valvules flap back against the adjacent walls, permitting free flow of fluid. If the pressure is higher on the normally downstream side, it forces the valvules to move toward the center, where they meet and block flow. The pulmonary valve controls the exit from the right ventricle; the aortic valve does the same for the left ventricle. Between the right atrium and its ventricle stands the tricuspid (or right atrioventricular) valve; between the left atrium and ventricle is the mitral (or left atrioventricular) valve. These latter two are the valves that the papillary muscles of the ventricles help keep closed when the ventricle squeezes.

A few other valves are almost always present, but their functioning isn't quite so critical. There's one between the inferior vena cava and the right atrium, although a few of us lack it as adults. As mentioned, no valve intervenes between superior vena cava and atrium, but there are small ones where the coronary veins return blood to the right atrium.

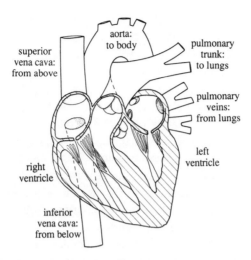

Figure 2.7. The heart, a somewhat diagrammatic cut-away viewed from the front.

Heart valves generally work well, although some incomplete sealing, with consequent backflow, isn't uncommon. The backflow frequently makes noise detectable with a stethoscope—it's called a *murmur*.

Passively operating one-way valves aren't anything very special—we use lots of them in our technology. Commercial and biological check valves achieve their identical result in quite different ways, though, ways that reflect one of the great divides between human and natural technology. All of the natural valves we've been considering (and most oth-

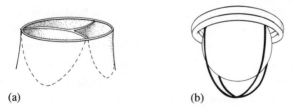

Figure 2.8. One way valves. (a) is a typical heart valve; (b) is a commercial ball-in-cage valve. The latter happens to be one designed to replace a heart valve. As shown, both permit flow from top to bottom and prevent flow from bottom to top.

ers as well) use a set of flexible flaps that bend forward and outward to open and that bend backward and inward to meet and close. The most common commercial valves use a ball or plate that can either be pushed forward to expose some channel around ball or plate, or pushed backward to close off another channel and stop backflow, as in Figure 2.8b. We just don't often design objects that function by repeated bending—our technology stands stolidly on stiff materials. Door hinges, for instance, use discrete solid elements sliding across one another rather than flexible structures that work by bending. We've no special aversion to unbound elements such as wheels, hinge pins, or the balls in the valves just mentioned. By contrast, nature's materials are more often flexible than stiff. Relative motion of parts is more often accomplished by bending than by sliding. Nature confidently uses joints that bend repeatedly for millions of operations, such as the inner edges of these heart valves. On the other hand, nature appears to abhor all but very tiny detached elements—in organisms everything is usually connected somewhere with everything else, not just in direct contact.

Mammals without Functional Lungs

At first, the image doesn't sound at all nice. Still, we were all that way, once, since a mammal must be born before it can breathe. The consequences for circulatory systems and for hearts in particular constitute what must be the most amazing of all the events attending a mammalian birth. Fetal blood, of course, gets its oxygen from the maternal blood, so mother acts like a surrogate set of lungs. The arrangement involves the close proximity of maternal and fetal capillaries in the placenta. Quite reasonably, only a little blood passes through the developing fetal lungs. But the normal mammalian circulation pattern mandates a pulmonary blood flow exactly equal in rate to that of the rest of the body combined, an enormous rate for a single organ. Obviously the fetus can't be using that normal pattern.

What happens is shown in Figure 2.9. Blood from the placenta enters the inferior vena cava of the fetus via the umbilical vein and passes, in normal fashion, to the right atrium. Most of this blood, however, now passes through an opening between the two atria and thus gets to

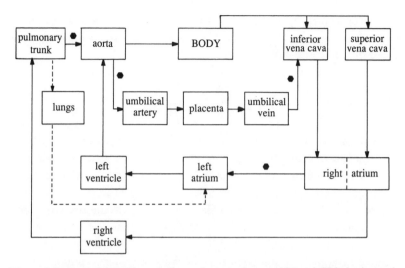

Figure 2.9. Something this complex only a mother could love! This is the path of the blood in a fetal mammal; you might compare this to the diagram of Figure 1.3, page 13. At or shortly after birth, the connections shown dashed open up, and those marked with hexagons close off.

the left side of the heart without going through the lungs. What of blood from the superior vena cava? It also goes to the right atrium, but most of it flows normally into the right ventricle. (The right atrium is doing two things at once, in a fine bit of fluid-mechanical sleight of hand.) Coming out of the right ventricle most of it passes, not to the lungs, but through an opening between the pulmonary trunk and the aorta. In effect, the four chambers are pumping as a parallel pair of two each, with upstream connections between the two atria and another connection just downstream of the ventricles.

Then comes birth. The lungs inflate, lowering their resistance to flow. As a result, blood from the pulmonary trunk goes freely to the lungs and returns in the pulmonary veins to the left atrium. As a result, pressure rises in the left atrium. At the same time, the umbilical vein begins to close off, and pressure drops in the right atrium. This shift in the relative pressures of the atria closes the valvular opening between them, and tissue eventually grows over the structure, usually sealing it for life. At the same time the duct between the pulmonary trunk and the aorta begins to close off, although some blood flows through it for a week or two. Voilà—the adult serial arrangement of half-hearts instead of the fetal parallel version.

And an Alternative World

This whole account of hearts has focused on mammals particularly and birds approximately. It turns on a set of structures with clear ancestral as well as functional equivalents in other vertebrates. One might ask, however, whether really different arrangements could work as well, about whether a designer with a fresh slate might have come up with something really distinctive. Insects, as we've already seen, get their oxygen distributed in a fundamentally different way, so different that one can't make easy point by point comparisons. Another group of animals, again quite out of the lineage that gave rise to vertebrates, has produced large and active creatures; looking at them (which will happen repeatedly as we go on) provides as nice a contrasting perspective on circulation as one could wish. These are squids, cuttlefish, and octopuses—the class of mollusks called *cephalopods*.

These cephalopods, it's a pleasure to report, have come up with the same scheme as the mammals and birds. It's the same, though, only in a purely functional sense—the components look quite a lot different and are spread around in far from corresponding places. The cephalopods have moved beyond the single hearts of their molluscan ancestors just as we have transcended those of our fishier vertebrate forebears. And, just as in our lineage, a second pumping system gets around the problem of making blood go through two exchangers, two capillary beds in sequence. But the cephalopods do the trick, not by partitioning a single contractile machine, but by the addition of a pair of booster hearts that force blood through their gills. In short, as shown in Figure 2.10, cephalopods have three hearts. The main, systemic heart receives blood from the gills and pumps it fore and aft through arteries to the tissues of the body as does our left half-heart. The two gill hearts, one on each side, receive blood from the body and send it to the gills. The main heart has not one but two atria, each passing along blood from one side of the animal.

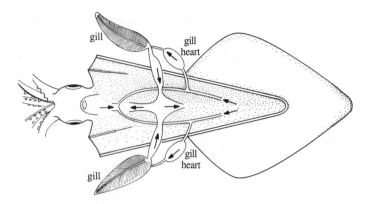

Figure 2.10. The circulatory arrangement of a squid (or an octopus). The figure is a bit diagrammatic—the outer mantle has been opened and the gills folded outward. The central chambers are the two atria and the ventricle of the main (systemic) heart.

In a way, the circulatory system of these cephalopods appears a little more logical than ours. The secondarily evolved pumps are clearly boosters serving the respiratory organs, and the system is bilaterally symmetrical. That means it can be divided by a plane into left and right mirror images, the symmetry both we and the cephalopods show on the outside. Our booster pump, as mentioned earlier, is the old main one of the fish; our external symmetry belies a heart asymmetrical in shape and asymmetrically located to the left of the midplane of the body. Again, though, no judgment is implied. My guess is that we're just the confused result of evolution both of a dual-pump system and of new gas-exchange organs, with lungs replacing gills in the transition to air-breathing. But although lungs do the old gill function, as we noted in the last chapter, they aren't made, evolutionarily or embryologically, from gills or the precursors of gills. Thus there's no reason to expect them to be hooked to the circulatory system with the equivalent vessels. By contrast cephalopods are now and have always been marine creatures; their gills aren't fundamentally different structures from those of less active mollusks such as clams.

The cephalopod arrangement and the molluscan system in general seems a more reasonable design in another respect, one probably reflecting a much earlier evolutionary event. In the basic vertebrate scheme the heart pumps blood to the gills, so in consequence the heart moves deoxygenated blood. So even a fish can't use the blood passing through its heart to supply the heart's muscle, and it needs (and has) auxiliary coronary arteries to do the job. A basic molluscan heart, on the other hand, pumps blood from the gills to the rest of the body. It thereby handles oxygenated blood and can supply its own metabolic need for oxygen from, as it were, the mainstream flow. Vertebrate hearts are fundamentally weird—cafeterias that send out for food for the employees.

Cooking with Heart

Back to the here and now with the promised guidance for cooking heart, either for what you have left over from your anatomical explorations or if you want to eat cheap red meat without worrying about your own heart and coronary arteries. (But don't try to cook embalmed heart!) The basic culinary problem isn't the muscle, it's the other main structural component of heart, a protein called *collagen*. The name means, from the Greek, "glue source"—"colla" as in "collage" and "gen" as in "genesis."

Collagen is the ultimate source of old-fashioned glue made from bones, skin (hide glue), fish offal, and other such food byproducts. It's also the source of gelatin. Within organisms, collagen is a fibrous, water-insoluble protein, perhaps the most common of all animal proteins. Its fibers are strong and flexible, but they are ropey rather than elastic or, put another way, not very stretchy. Thus when pulled on by contracting muscles they convey that shortening to bone, skin, or some other muscle rather than themselves stretching. Contract your calf muscles and you stand on tip-toes—your collagenous Achilles tendons pull on the heels. Literally laced with collagen are boneless muscular organs—the outer mantle of a squid, many tongues, and, of present interest, hearts.

This ropey material, collagen, is the very essence of toughness. Roast or grill heart, and you've a real jaw-wearying product. You have to do a mild version of what the gluemakers do, solubilizing the fibers by chopping up the long molecules. The process is termed *acid hydrolysis* since it amounts to inserting water molecules at break points under acidic conditions. A long, cool incubation followed by a shorter hot treatment works fine—in short, what one ordinarily does for tough meat. The choice of acids is more cultural than critical, with vinegar, lemon juice, and tomato sauce as common contenders for the task. The following is a recipe derivative of a south Indian stew called a *vindaloo;* an alternative is a German sauerbraten. (Despite the traditionally fiery character of vindaloos, you may be pleased to know that the red pepper isn't critical to the tenderization process. Anyway, this is a fairly non-corrosive one.)

1/2 tsp crushed red pepper

10–20 garlic cloves, sliced

1 tsp ground cumin

1 tsp powdered mustard

1 tsp ground turmeric

2 cubic inches ginger root, chopped (or 2 tsp ground ginger)

1 tbsp lemon pulp

1 tbsp sugar

4 tbsp poppy seeds

3/4 cup vinegar

2 lbs heart (ventricle), cut into small cubes

2 tbsp vegetable oil

4 cloves

1 onion, half-rings

1/2 cup tomato sauce

Blend pepper, garlic, cumin, mustard, turmeric, ginger, lemon pulp, sugar, salt, poppy seeds, and vinegar at high speed until homogeneous. Mix with meat in a bowl and marinate overnight. Heat oil in a Dutch oven, add cloves, and fry a few minutes. Then add onions and fry until they're soft. Cool a bit to lessen the danger of sticking or splattering and add the marinated mix. Cook slowly, covered, for an hour or two. Add tomato sauce and cook until meat is tender, usually about another hour. Uncover and cook off extra liquid if necessary.

Sources of Preserved Hearts

All of these vendors offer embalmed cow or calf, sheep or lamb, and pig hearts. They all have horribly expensive anatomical models, but Fisher and Wards have less costly ones as well. For nonfondlers NASCO has a reasonably priced sheep heart sliced open and embedded in plastic, and Wards has freeze-dried whole and bisected (as opposed to dissected) sheep hearts. Addresses:

Carolina Biological Supply Co.
Burlington, NC 27215
  919-584-0381

NASCO
901 Janesville Ave, Box 901
Ft. Atkinson, WI 53538
  414-563-2446

Fisher Scientific Company
Educational Materials Division
4901 W. LeMoyne Street
Chicago, IL 60651
  312-378-7770

Ward's Natural Science
Establishment
P.O. Box 92912
Rochester, NY 14692
  716-359-2502

A good source for inexpensive models (but whose catalog may be dangerous to your financial security) is

Edmund Scientific Co.
101 E. Glouchester Pike
Barrington, NJ 08007
  609-573-6259

Notes

1. For lack of day-to-day utility it may be only a distant memory, but you probably once learned that the volume of a sphere was four-thirds times pi times the cube of its radius.
2. Just air if you have an engine that uses fuel injection.
3. According to the U.S. Department of Agriculture's book, *Composition of Food*, heart, even with a little superficial fat layer, has only 3.6 percent fat. For

comparison, flank steak has 5.7 percent, round has 12.3 percent, and the really fancy cuts around 35 percent. The book, from the Superintendent of Documents, Printing Office, Washington, D.C. 20402, is something no home should be without.

4. You can guess that I'm assuming a bipedal, upright animal such as one of us. I'm trying to avoid using *dorsal, ventral, sagittal,* and such other directional descriptors. Anyhow, one usually draws even dissected quadrupeds as if the head were up and the tail down.

5. Not really as difficult as it sounds—with fruit-flies one can tolerate a very large failure rate. More impressive are the people who do delicate work on other people, where the Damocles' sword of a malpractice suit may be more than a match for a scalpel.

3 *Getting There*

Before penetrating further into the mechanics of circulation, let's take a brief look at how we got to know about a few of these things. While this isn't a book about the history of science, there's no denying that it's about hard-won knowledge gained by some impressive people. Giving them a little attention gives me a chance to say some things about how science actually happens. The process is a great deal less steady and stodgy and far more fraught with human foibles than one might judge from its results.

Discovering the Circulation—William Harvey

Beyond any doubt, the classic figure in the circulation story is William Harvey (1578–1657). We attribute to him the discovery of the circulation of blood rather the way evolution by natural selection is attributed to Darwin and Wallace or the modern notion of particulate inheritance to Gregor Mendel. Each of these cases has an element of what might be called "personification"—we like our history replete with heroes and unsullied by pedantic and pedestrian problems of priority. But these are genuine giants, fine heroes, and Harvey isn't the least of them. Claiming circulation for him is certainly less than the expected exaggeration or oversimplification. Beyond that specific discovery, he marks the edge of our side of an intellectual divide in functional biology—his position is analogous to that of, I suppose, Vesalius in anatomy or Thucydides in history. To read the works of his predecessors is to step back into an earlier world of science in which the reasoning is unfamiliar and the view is muddied with pervasive misconceptions. They're clearly bright and imaginative workers, but in some way I find hard to pin down, they aren't quite of our genre. By contrast, Harvey is not just one of us, but is still a good model for the best of us.

The general picture of blood flow prior to Harvey was one that now strikes us as most peculiar. Blood was manufactured in the liver, the largest of the visceral organs and one penetrated by large blood vessels. Fresh liver is, indeed, pretty bloody material. Blood was shipped out to the rest of the body in the veins to provide nourishment. The heart generated the heat of the body and was refreshed or cooled by air coming from the lungs; blood charged up with the resulting "vital spirits" went out in the arteries. Now this is neither unreasonable nor illogical, particularly in an era innocent of any notion of metabolic or even combustion chemistry. One can emphasize the point by modernizing the picture, recognizing that the liver is the body's short-term energy store and exports glucose, that the lungs similarly export dissolved oxygen, and that the cells of the body survive by combining the two. The trouble is that, however reasonable and logical, it's fundamentally and irremediably wrong as a view of circulation. It's like the Lamarckian scheme (for accentuation of a feature as a direct consequence of increased use of it) for evolution—not unreasonable, just unreal.

Before Harvey, the authority for over 1000 years had been the Greek physician Galen, who made magnificent discoveries, but who managed to head posterity off in the wrong direction. Working from dissections of dead material, he got the mechanics magnificently muddled. Galen wasn't convinced that the heart was muscular, since it would move without our will and even after its nerves were completely severed—fine facts, wrong inference. Arteries expanded and contracted, as you can easily feel; it wasn't clear that these were unpowered, passive changes responding to a central pump. Capillaries were invisible and not even suspected. So the system (oversimplifying, admittedly) was a reciprocating one. The heart got smaller and the blood vessels larger as blood moved outward; then the blood vessels got smaller and the heart got bigger, nicely accounting for relative volume changes. Again, the system wasn't entirely illogical—we tend to heap undeserved scorn on those responsible for the mistakes of the past even when their interpretations were completely reasonable given the facts they had available.

Anyhow, in a mere fifty-two pages of prose (*On the Motion of the Heart and Blood in Animals*, 1628), Harvey makes a revolution (an unusually apt term for discovering *circulation*). He marshalls the facts, new and old, beginning by dismantling the accepted, if by now a bit shopworn, Galenic paradigm. The logic, especially given the complexity of the system, is impressive.

1. Arteries are reinforced against higher pressure inside, not outside, which is quite opposite the arrangement found in the tracheal air pipes of the lungs (or, to use a modern analog, vacuum cleaner hoses). When cut, blood spurts out at the phase when an artery gets larger, not smaller. Where an artery has the thinnest wall, at an aneurysm,[1] its pulsation is greatest, not least. Thus arteries don't actively suck blood from the heart, and the heart must consequently propel the blood into the arterial system.

2. Arteries never have air inside, nor does the left ventricle, which draws blood from the lungs and sends it to the arteries. "Good God," says Harvey, a man of fine passion, "how should the mitral valves prevent the backflow of air and not of blood?" And the pulmonary veins, purported to carry air into the heart are even less like good airways than the arteries . . . "But I should like to be informed why, if the pulmonary vein were destined for conveying air, it has the structure of a blood vessel?"

3. To be sent out, vitalized, in the arteries, blood (originally from the liver and thus entering the heart through the inferior vena cava) would have to get from the right side of the heart to the left. The old scheme had another role for the lungs, so pores were postulated in the muscular wall between right and left ventricles. But the necessary pores had evaded even the sharp eyes of Vesalius, so Harvey has support for his scorn: "by Hercules, no such pores can be demonstrated, not in fact do any such exist . . . But even supposing there were foramina or pores in this

situation, how could one of the ventricles extract anything from the other . . . when we see that both ventricles contract and dilate simultaneously?" Finally, he points out that the connection exists in the fetus; if it were required in the adult, what would be less logical than to shut it down right after birth?

He then builds his new edifice, using both his own findings and those of his predecessors—for the latter carefully separating facts from errors and speculations. First, he argues persuasively that heart is proper muscle and its active motion is contraction, in which it gets firm and tense, not dilation, when it gets softer. And he explains how a muscle can be made to squeeze a chamber, and reminds us of his demonstration that the arterial pulse is passive, a mere reflection of the action of the heart. Turning again to the heart, he carefully describes the sequential contractions of auricles and ventricles and the action of the valves between them. He describes the copious pulmonary blood flow and the way in which blood is shunted around the lungs in the embryo.

At this point he brings the pieces together and describes the pulmonary circulation—how blood can only get from right ventricle to left by passing through the lungs, postulating only the "hidden porosities of the lungs and the minute inosculations of vessels." Blood moves from veins to arteries by way of lungs, and he notes that right ventricles that might propel that passage are features only of animals that have lungs. The notion of overall circulation is then finally introduced, with the first understated mention of the idea that just as blood got from the veins to the arteries through the lungs, it must "somehow find its way from the arteries into the veins." That, of course, makes explicit the second undescribed element, the smallest vessels of the systemic circulation. Recognition of both unknowns followed 30 years later, when the first great microanatomist, Marcello Malpighi, discovered the capillaries, initially in the lungs of frogs, and thereafter elsewhere.[2]

Two substantive arguments are then brought to bear on the overall proposition. The first (Harvey's order) is quantitative. He estimates the volume of the heart and the fraction that might be pumped into the aorta; multiplying the two gives him the volume pumped per beat. Multiplying the latter by the rate at which the heart beats gives the volume pumped per unit time—cardiac output. Comparison with the weight of blood in entire animals (sheep and dogs, specifically) reveals that the entire mass would pass in less than half an hour. As he triumphantly proclaims, ". . . a vastly greater amount would still be thrown into the arteries and whole body than could possibly be supplied by the food consumed. It could be furnished in no other way than by making a circuit and returning."

The second argument rests on the demonstration that veins have valves that permit only heartward flow of blood. We'll return to the demonstration of the valves in Chapter 7 since Harvey's way of showing them isn't mere history. It hasn't been improved upon and can be done on oneself. The remainder of his book is largely devoted to lacing the idea of circulation into the existing body of physiological and medical knowledge, showing that it is either consistent with, or positively supported by, a host of other observations.[3]

* * *

What was so special about Harvey's approach to his problem? I'm impressed with several quite independent items. For one thing, he based his conclusions on functional experiments and observations on living animals, thereby recognizing the errors that had accumulated as a result of reasoning from anatomical studies on fresh but nonetheless nonliving material. For another, his specific experiments were cleverly conceived and executed. In addition, he made good use of the perspective gained by looking at blood flow and the associated equipment in fish, frogs, snakes, shrimp, clams, and other animals. That certainly warms the heart of a biologist!

More impressive, though, is the way he brought measurements and quantitative arguments to bear on his problem. The argument against pores in the interventricular wall was in part based on the very large volume that would have to pass through those hypothetical pores in a thick and dense wall. The major piece of quantification, of course, was his calculation of cardiac output and comparison of it with the mass of the blood. It was a quantitative argument in support of an essentially qualitative proposition, the *circulatory* path of the blood. The approach characterizes most of modern science. We don't collect all those numbers merely because we like numbers. Most often we're trying to do what Harvey did in order to evaluate a hypothesis that isn't fundamentally quantitative but still must pass the test of quantitative feasibility.

Still more impressive is the way Harvey rose above the prevailing way of thinking about the system. He combined a thorough understanding of what his predecessors had found out with an impressive detachment from their conceptual schemes. Any practicing scientist can tell you what a tough combination that is to manage. Most of the time most of science is done within, or perhaps one ought better say fitted into, an existing general picture or paradigm. Details are accumulated, errors are discarded, and the focus of the picture is gradually sharpened. Sometimes, though, new data doesn't sharpen the picture—it merely complicates it. What seemed simple and direct can gradually get a bit jumbled, and a tidy theory takes on the character of a bumpy, uneven synthesis. Of course, the way out is a new and more powerful paradigm; but that's easier said than done. For one thing, it's not often clear in actual practice exactly when the time has come to start afresh. Worse, all those details have to be dealt with in the new view; and that means that whoever is to generate the new paradigm must meet two nearly antithetical requirements. The person needs rather good (but certainly not perfect) familiarity with the details, both old and new. But equally important, the person must be conceptually independent of the old paradigm.

No better example exists of someone who knows the old but isn't enthralled by it than Harvey. Venous valves, for instance, were discovered by Harvey's teacher, Hieronymus Fabricius. Fabricius, however, imagined that they acted as brakes to prevent blood from moving too rapidly to the appendages—he managed to fit them, if a bit uncomfortably, into the old paradigm. Harvey's is the classic example of a new paradigm in physiology, at once more simple and more powerful than the old in just the way the Copernican sun-centered solar system proved both simpler and more powerful than the Ptolemaic earth-centered one. It's too easy to forget that a major aspiration of science is describing the world with the minimum number of maximally predictive statements or organizing principles. Facts are certainly necessary in science, but to think that science mainly advances by the patient accumulation of facts or data is a very great (and very common) misconception.

A parenthetical note on Harvey's quantification. An article I read some time ago took Harvey to task for using bad data in calculating cardiac output. That's unfair on two accounts. First, in context it's amply clear that the calculation is based on a kind of "worst case" scenario. Using a heart rate of 33 per minute while noting that rates four times higher are common is certainly no worse than taking 72 as an implied standard. The assumption that the heart holds two ounces isn't a grotesque underestimate. (Five ounces, the figure I cited in Chapter 1, might be a better average.) Assuming that an eighth of the volume moves out with each stroke is a major underestimate, but Harvey makes it clear that in the absence of better data, the most conservative figure is the one to use. The result of these underestimates is that cardiac output is given as less than a tenth the value we now report. If even that low value suffices to make the argument for circulation, so much the better—a point Harvey quite explicitly states.

This point about quantification is no minor matter. A pervasive mythology holds that when doing science one should be as accurate as possible. If we adhered to that notion, we'd make no progress at all. Reality involves a more rational if more complex standard. One figures out at the start just how accurate the data has to be to test the hypothesis, to answer the question. And then one wastes no time being any more accurate than that. The trick is to take accuracy not as some metaphysical Grail, but as something you quantify, specify, and keep track of—the accuracy of measurements should be dealt with in as systematic a way as the measurements themselves. Strictly speaking, no datum in science is of any use unless one can specify its accuracy; that is, unless one can say quantitatively just what confidence it merits. Certainly Harvey's data win no prizes for accuracy. The critical second point is that science awards no prizes for accuracy per se, and his data were amply adequate for overturning the Galenic applecart.

And a point about the rationality of biological design. Organisms are not random accumulations of matter but appear to be quite well attuned to their individual niches in the world. That they are "well designed" can be argued either with reference to a benevolent and omnipotent creator or to evolution by natural selection. Both referents have drawbacks, but either can serve as a rough and ready rationalization. If some aspect of an organism looks badly designed for its function, the best working hypothesis is that its function is badly understood. Harvey had a nice sense of this peculiarly biological principle, as we saw in his comments on the closure of the interventricular connection after birth, on the venous valves, and on the design of the pulmonary veins as blood vessels, not air ducts.

Controlling the Heart—Otto Loewi

Harvey's is a fine account, but at least by a sin of omission it is thoroughly misleading. For the purpose of writing an article or a book a post facto logic is created that typically conveys little about the real origin of a concept or the chronology of verification. No deception is intended, and the scientific literature is much the more manageable for the practice. Admittedly it's tacitly pretentious, intimidating to neophytes, and misleading to nonscientists; but we've no patience for Proust, stream of consciousness, or strict history when we're dealing with the contemporary volume of scientific literature. Only rarely does someone let down the hair and baldly relate things as they really happened. Such accounts may be both enjoyable and revealing; but when I have to work through dozens of papers in a few days, I've no time for introspection, blind alleys, or verbal splendor.

The real trouble is that an organized account is by its very nature a rational account, so it's a little tricky to talk about the irrational in the same format. Yet there's quite clearly a major element of irrationality in scientific progress, the central item of which is the origin of ideas. Harvey, for instance, gives no credible hint about where he got the idea of the circulation nor about when or how he began to have serious doubts about the prevailing paradigm. The cold logic with which an idea is finally presented certainly doesn't reflect the heat of conception. Robert Root-Bernstein has written an account of this mysterious business of idea generation that I find penetrating, iconoclastic, and thoroughly enjoyable. For now we'll talk about one person and one discovery—the discovery of the neurochemical control of the heart, in 1921, by Otto Loewi (1873–1961). At that time he was professor of pharmacology at Graz, Austria; in 1936 he shared the Nobel Prize for Physiology or Medicine with Sir Henry Dale for this and subsequent work.[4]

Recall that Galen didn't recognize that a heart was made of muscle. The muscle that he was familiar with, what we call skeletal muscle, doesn't speak unless spoken to—nervous stimulation is needed to trigger contraction. If removed from the body it lies there, limp and unimpressive, except as meat. A vertebrate heart, in sharp contrast, is intrinsically active. It's fully capable of beating after removal from an animal, at least for a while, and in the body it will beat indefinitely after all nerves leading to it are severed. To put the matter rather grossly, an electrically dead brain is completely compatible with a fully functional heart. Forensic medicine has not been comfortable with the phenomenon.

Curiously, hearts with intrinsic activity are not universal. These so-called myogenic hearts (rhythm *gen*erated in muscle, or *myo*cardium) occur in vertebrates and mollusks, but the hearts of many arthropods such as spiders and crustaceans have "neurogenic" hearts that work only with a functional nerve supply. From what we know about how nerves and muscles work, an intrinsic rhythmicity in a muscle seems a fairly ordinary thing, a minor specialization of a normally quiescent device. In fact, lots of nerves are also spontaneously active, discharging impulses in the absence of stimulation from any external source such as other nerves.

Our hearts, along with those of most other vertebrates, receive stimuli from the central nervous system by way of two nerves. Stimulation through one nerve (called a *depressor*) reduces the rate at which the heart beats; stimulation by way of the other nerve (the *accelerator*) increases the frequency of heartbeat. That much was known to Loewi before his now famous experiment. He was interested in the mechanism by which a nerve affected a muscle; for him heart was just one case. Years later, in one of a series of lectures, he related the actual events:

In the night of Easter Sunday, 1921, I awoke, turned on the light, and jotted down a few notes on a tiny slip of paper. Then I fell asleep again. It occurred to me at six o'clock in the morning that during the night I had written down something most important, but I was unable to decipher the scrawl. That Sunday was the most desperate day in my whole scientific life. During the next night, however, I awoke again, at three o'clock, and I remembered what it was. This time I did not take any risk; I got up immediately, went to the laboratory, made the experiment on the frog's heart . . . and at five o'clock the chemical transmission of nervous impulses was conclusively proved.

What he had done was to remove the hearts from two frogs (frogs, again), maintaining them in salt solutions that mimicked intercellular body fluid. The first heart had been taken out along with its depressor nerve attached, and that nerve was repeatedly stimulated, slowing the heart. The second heart was then transferred to the solution that had bathed the first, and it responded by slowing down in just the same manner without any neural stimulation at all. Stimulation of the nerve had not only slowed the first heart, but in the process it had released a substance with a depressor action into the solution. It was later shown that one could do the same thing with the accelerator nerve, making a second heart speed up when exposed to the solution that bathed the first. Two substances were involved, and they were later identified as acetylcholine, the depressor, and adrenalin, the accelerator. These turn out to be chemical connectors that link other neuromuscular junctions and many nerve–nerve junctions as well.

(Substances that block the cardio-accelerating action of adrenalin— beta blockers—are commonly given to people who have heart problems or high blood pressure. They're quite useful but, like most medications, they can do more things than we'd like. The trouble in this case is that entirely too many organs have so-called beta-adrenergic receptors. Multiple targets, whose physiological utility for this case will be argued shortly, commonly constitute a therapeutic nuisance.)

Incidentally, Loewi was not just clever, he was on this occasion lucky. Had he used blood instead of salt solutions to maintain the hearts, no effect would have been seen—blood contains an enzyme that inactivates acetylcholine. In addition, the experiment turns out to work more satisfactorily with frogs than with most other animals; it's in part a matter of how much of the substance leaks out of the heart into the blood.

But I mean to dwell not on the luck but on the irrational element. Where did the idea for the definitive experiment really originate? Loewi himself comments that "in 1903 I had already expressed the view that, as certain chemicals act exactly like stimulation of certain nerves, it may be that these nerves, in their turn, act by liberating chemical substances. This I had entirely forgotten." And in the same lecture he explicitly recognizes the irrational, intuitive element: "One thing, however, is clear: though the essence of art and science is different insofar as art appeals to our emotion, science to our reasoning, there exists a huge body of evidence that intuition is at the basis of creation in science as well as in art."

I will leave any more penetrating analysis to psychologists and philosophers. What worries me, though, is the thought that we might now be actively preventing the kind of thing Loewi did. It's bad enough that it usually takes multiple attempts with lengthy applications and at best nearly a year's delay to get funding for some new project. It's also bad enough that it's especially hard to get funds for something regarded as a long shot. One can often do a critical experiment using material on hand. But what if Loewi had to wait months after describing the proposed (and most peculiar) protocol for uncertain approval by some bureaucratic body that regulated the use of animals for experimentation?

I'm fond of animals, including frogs; and I've never seen any sign among the biologists I've known of the sadistic mentality that might derive gratification from killing or injuring animals. Most often I've noticed an attitude reflecting unpleasant necessity. Perhaps, to prevent unnecessary hardship to experimental animals, any scientist who directs work involving animals ought to make a practice of doing some of the experimental protocols personally. Such contact might both improve the overall quality of the science and ensure the most sober and judicious use of experimental animals. (But, then, I'm the kind of fanatic who believes that anyone who supports capital punishment should be willing to act as executioner.) It seems to me that people opposed to any use of animals in biological science are unrealistic, sanctimonious, and (whether by ignorance or intent I couldn't say) arguing against doing biology at all. Computers may be wonderful for exploring the consequences of a model or for analyzing data, but the organisms themselves are the ultimate reality against which any hypothesis must finally be tested.

Making the Heart Beat: The Rest of the Story

Merely declaring our hearts myogenic would be to omit an interesting story. An entire heart isn't intrinsically active. Rather, only a few specific areas of the heart have the capacity for spontaneous muscle contraction. In each, a contraction is normally followed by a period of inactivity. Except at the very start of that quiescent period, however, the muscle can be made to contract by stimulation from elsewhere. What results is yet another really neat scheme. Whichever area has the fastest rhythm sets the pace for the whole organ. If that portion is in any way incapacitated, then the next most rapid area automatically acts as pacemaker. If no area is rapid enough, an artificial pacemaker can be installed, and it quite automatically takes over and suppresses any other.

The normal pacemaker is a group (*node*) of cells in the wall of the right atrium near where the superior vena cava leads into it. From there a wave of muscle contraction rapidly spreads over the two atria, so their entire walls give a coherent squeeze. At the same time, the electrical signal from the original nodal cells travels over several specific paths (again a comforting bit of redundancy) to another group of slightly special trigger cells. These latter are located at the base of the right atrium, where it contacts the ventricles. As a result, the ventricles get triggered about a tenth of a second after the start of atrial contraction. Ventricles are large, and their simultaneous and coherent contraction is assured by a specific group of rapidly conducting pathways. These carry the triggering signal from the second set of nodal cells through the rest of the muscle mass of the ventricles. The overall result is a properly phased sequence of squeezes—first the atria and then the ventricles.

The substances, acetylcholine and adrenalin, that slow down and speed up the heart exert their effects on the first set of nodal cells. Acetylcholine, because an enzyme that breaks it down, doesn't get carried by the blood. Adrenalin, on the other hand, can be and is disseminated in the bloodstream. Thus the heart can be speeded up by the adrenalin from either of two sources. Adrenalin may be released by the end of the accelerator nerve coming from the central nervous system; alternatively, it may be produced by a gland, the adrenal, on the surface of the kidneys, and released into the blood. If the former happens, then speeding of the heart can be the sole effect; if the latter occurs, then the adrenalin triggers a whole spectrum of reactions. This set of reactions is appropriate to fear, anger, or excitement—increased blood pressure and blood flow to the heart, lungs, and body muscle; release of sugar into the blood; and others. At the same time less blood flows to the skin and digestive system—fear can make your skin cold and give you cramps. The whole scenario triggered by release of adrenalin into the blood is commonly called the "fight or flight syndrome."

Is adrenalin a hormone or a neurochemical? We avoid the distinction by calling it a neurohormone. What's more important is the lack of clear boundary between nervous and endocrine systems; the distinction is ours, so why should nature care? She's[5] not being capricious, but merely remaining opportunistic. In a larger sense, the presumption of tidy boundaries represents a misreading of what must be the central message of physiology, the complex and comprehensive integration of our functional components.

Notes

1. Aneurysms will get proper attention in Chapter 7.

2. We really ought to say a word on behalf of frogs, involuntary martyrs over hundreds of years in the cause of advancing our knowledge about circulatory physiology. As Malpighi comments in a context implying suffering on both sides, "For the loosening of these knots I have destroyed almost the entire race of frogs . . ."

3. Beyond arguing the case for a revolutionary view, *On the Motion of the Heart and Blood in Animals* is also an unusually enjoyable account if one knows just a little of the relevant physiology. I recommend it and assure any prospective reader that the present book provides a more than ample background.

4. Loewi moved from German Austria to the United States in the late 1930s, part of the sad exodus that marked the practical end of the great German-language scientific tradition.

5. Even Harvey refers to nature as "she." I'm not sure of the origin of the practice (the Oxford English Dictionary traces it back to Chaucer, at least) but I approve as a matter of conviction and not just convention. It recognizes that

biology is intrinsically and inescapably a sexist subject, specifically, feminist. Males are an afterthought, a side issue that takes some explaining by the evolutionary theorists. All-female species exist; no all-male species will ever be discovered! We quite properly call the result of a cell division "daughter" cells. Species in which males are larger are far less common than ones in which females are larger. And so on—Eve invents Adam in our biological cosmology.

4 Pressure

At regular intervals I am subjected to that ritual taking of the pressure of my blood without which no medical examination can claim respectability. The result, these days, is usually a pair of numbers, approximately 110 and 70, and some mild murmur of approval by the examiner. Someone is apparently convinced that pressure is a significant variable in cardiovascular operation, or at least that an abnormal pressure signals something pathological. (Abnormal and pathological, we and our physicians must remember, are not exactly synonymous.) I've no disagreement—indeed, we can't talk much more about circulation without getting a little more explicit about pressure. It's handiest, though, to approach pressure from afar rather than going directly into any formal definition.

Pressure, Viscosity, and Flow in Pipes

Neither air nor water nor blood will flow through a pipe without some sort of push or pull—in short, without provision of force. The statement is self-evident, intuitively obvious from all of our everyday experience. In fact, this is one of those cases where reality seems simple until the relevant body of underlying physics is examined. According to Newton's first law, a body (which includes a body of fluid) will remain at rest unless some force is applied to move it. No trouble so far.

But furthermore, still by that first law, a body will remain in steady motion with no further application of force once it is brought up to speed. The implication is clear enough—once the heart gets the blood going, its labor is done. What reconciles the first law with the sustained need for a heart is friction, not exactly the friction of a sled pulled across a patch of bare pavement, but its fluid mechanical analog.

There's a strange phenomenon, called without salacious intent the *no-slip condition*, that must be brought up at this point. When any fluid flows across any solid surface the fluid at the surface stays put—an infinitely thin layer of fluid (ignoring molecular phenomena) stays with the solid, not with the main body of the fluid. If you were to make a set of measurements of the speed of flow at different points, gradually approaching the solid surface, you'd find that flow slows down and reaches zero speed just at the surface, as shown in Figure 4.1. In short, the fluid at the surface does not "slip" along the surface. The phenomenon underlies several ordinary observations. The blades of a household fan accumulate dust and grime whereas one might expect the high wind speeds to blow them clean; it certainly looks as if those winds don't quite reach the surface. Running water over a dirty plate is far less effective in cleaning it than a swipe or two with a dishcloth; maybe again the *flow* somehow doesn't make proper contact with the surface even though the *fluid* certainly does.[1]

Since the main body of fluid moves freely, a region exists near the surface in which (as in the figure) the speed of flow varies with the distance from that surface. It helps to envision the fluid as a pile of thin layers lying on the surface, rather like the pages of manuscript here on my desk. If I were to push with uniform force everywhere across one side of the pile, the pages would move; but the upper ones would move either faster or more easily, and the lower ones more slowly or with more difficulty. If the desk's surface were sufficiently rough (sandpaper, glued on, would do) then the lowest page wouldn't move at all. The salient point is that pages move across each other, with attendant friction. Much the same thing happens in the fluid flowing across a solid surface. The layers might be ultimately fictitious or (much the same thing) infinitely thin, but the same sort of gradient of speeds occurs together with the same sort of interlamellar stickiness. The same requirement exists for a force to keep things going, to counteract the

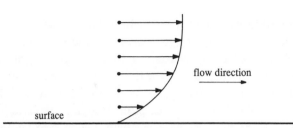

Figure 4.1. How the speed of flow varies near a surface along which a fluid flows. The lengths of the arrows represent the flow speeds at the points from which the arrows emerge.

same frictional tax. This friction, though, differs in two ways from that between solids. First, the friction, while caused by the presence of the surface, is developed entirely within the fluid instead of between solid surfaces or between solid and nearest fluid. Second, you can't, unfortunately, lubricate yourself around the problem.

Interlamellar stickiness—if it brings to mind taffy, you're thinking the right way for the occasion. The proper name for the property is *viscosity*; it's a property of all fluids from gases, with very low values, to glasses, where it is so high we treat them as solids over ordinary lifetimes.[2] Viscosity is resistance to deforming *motion;* it is quite distinct from the *stiffness* of solids, their resistance to the deformation itself. A fluid has no preferred shape—the coffee fits any cup, and when poured out it shows no memory of the shape of cup or pot. It does, however, care about how fast it's asked to change shape. To get a faster change more force must be applied, more work needs to be done, more energy has to be invested. And the more viscous the fluid, the more force is needed to get, not the same deformation, but the same speed of deformation. Again, a fluid cares only how fast it's deformed, by contrast with a solid, which, with some shape, fights forcefully against the deformation itself. The solid snaps back toward the original shape when the deforming force is removed;[3] the fluid does not.

Back to flow through pipes. Since pipes inevitably have walls, fluid is as inevitably flowing somewhere in the vicinity of a solid surface. Just as inevitably, viscosity exerts its pernicious frictional effects, and it takes a force to keep a fluid moving. Of course, that's the reason why the heart has to keep working. To proceed further, though, we have to move into an intuitively awkward realm. It's no trick at all to push with a force on a solid, to give an awl your all; nor is there conceptual trouble when pulling a plug. By contrast, pushing or pulling on a fluid requires a more devious approach—the fluid must be appropriately contained and the piston or turbine set to work. Pushing on a solid ordinarily propels it in the direction of the push. Pushing on a fluid propels it in any direction the fluid can get away with, which is to say that the container, as much as the push of the piston, determines the direction of flow. So instead of talking about the force we apply to a fluid, we speak of the *pressure* put on it.

In doing so, we bear in mind two peculiarities. First, pressure is defined as force divided by area. A push or pull of 1 pound on a piston (a large hypodermic syringe, if you can't envision pistons) with a plunger 1 square inch of surface corresponds to a pressure of 1 pound per square inch (1 PSI, sometimes). If the force is a newton[4] and the area a square meter, then the pressure is a newton per square meter, also called a *pascal* (Pa). Second, as mentioned, force has distinct direction, while pressure is omnidirectional and just positive (push) or negative (pull—much rarer and possible only in liquids, not gases). Apply pressure to the fluid in a container, and it will consequently squirt out equally well from an opening facing any direction. The latter is a real convenience, whether hooking up blood vessels or connecting the hydraulic lines from the master cylinders to the brakes on a car.

Measuring Pressures

There are still more quirks and queerness that can't be swept under the rug. Those numbers you're given while sitting with cuff on arm—their units, occasionally mentioned, are *millimeters of mercury* (abbreviated "mm Hg"). These correspond nicely to an up and down shift of the quicksilver in the U-shaped glass tube to which you are connected by a hose (Figure 4.2). No correspondence to force per area is self-evident. Nor are these the only pressure units in everyday use. The barometric pressure, portentous of weather changes, may be given as 29.93 inches of mercury—again, queerly, the height of a column of dense liquid. We've made the pressure do something and then just recorded the result. In

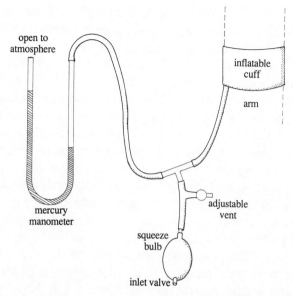

Figure 4.2. A mercury-filled manometer being used to measure the air pressure in an inflated cuff. The cuff is squeezing an upper arm, working in a way that will be explaned shortly.

these cases we make the pressure press upward on a column of the densest ordinary liquid, mercury, and we just report how far upward the column goes. One can imagine the pressure as an upward push on a (frictionless) piston at the bottom of a tube. Frictionless pistons are not easy to come by, which is where the U-tube comes in—pressure is applied to one side, which descends. The mercury has nowhere to go but up the other side, and we measure the difference between the heights of the two columns. That height difference is what the pressure is sustaining against the ever-lurking gravitational force. The instrument, by the way, is called a *manometer;* for taking blood pressures it's a *sphygmomanometer;* its meteorological equivalent is a *barometer.* The old Greeks once again supply roots: for *mano,* from the word for loose or spare; for *baro,* from the one for weight; for *sphygmo,* from the one for pulsation.

Conversion to sensible units of pressure is straightforward. One just looks up the conversion factors in a table and does the multiplication or division. For instance, multiplying millimeters of mercury by 133 gives pressure in pascals; multiplying inches of mercury by 3378 gives pascals as well; dividing millimeters of mercury by 51.7 gives pounds per square inch. (Reverse any of these operations by switching the words *multiply* and *divide*.)

We live, of course, immersed in an atmosphere that exerts its pressure on ourselves and on everything around us—in a region of gas extending a few tens of miles above the solid and liquid portion of our earth. Our gaseous atmosphere is made of real matter, which has real mass, which is therefore attracted toward the earth's center, and which consequently develops pressure in no way different from that of water on a diver. This atmospheric pressure normally exerts only subtle effects, mainly because for ordinary objects it's counterbalanced by equal pressures pushing outward from within. No particular magic is needed to achieve a perfect balance between inward and outward pushes—the inward press of the atmosphere squeezes an object until the outward pressure gets high enough for the two to be equal. If the object (water, flesh, or most solids) is resistant to squeezing (which is to say that it is nearly incompressible), then the volume change isn't normally noticed.

We do notice effects of atmospheric pressure when it changes, something that most commonly results from an ascent or descent since atmospheric pressure changes markedly with altitude. We notice it in particular when the air chamber inboard of our eardrums is insufficiently vented to the outside—when a cold results in plugging the Eustachian tubes connecting middle ear to nasal passages. An ascent reduces the outside pressure, and the air in the middle ear is free to expand a bit, causing, if unvented, an unpleasant outward bulge of the eardrum. As you may have noticed, the Eustachian tubes are open only during swallowing or yawning (but sometimes, painfully, not even then).

Atmospheric pressure isn't trivial. At sea level the average pressure is about 14.7 pounds per square inch (29.9 inches or 760 millimeters of mercury) with minor variations that, as mentioned, correlate with changes in the weather. That's about half the pressure to which you keep your automobile tires inflated. Several years ago I did some field work at the biological station at Bermuda. All of the critical and irreplaceable little things needed for the work got packed into a military surplus aluminum footlocker that had a nice waterproof gasket under its lid—a fine item worth far more than the $10 it cost (me, not the Pentagon). As it was off-loaded I noticed that the top was slightly bowed inward, and I wondered just what could have been dropped on it. At Bermudian customs the trunk persisted in remaining closed despite the combined efforts of me, my son, and the customs inspector. I finally wised up to what had happened. Aircraft are pressurized, clearly a requirement if ordinary folks are to be carried 6 or 8 miles above the solid earth. But they are pressurized to a pressure only about two-thirds that of the sea-level atmosphere. Going up, the trunk lid must have been pushed upward by expansion of the air inside and must have allowed some air to leak out. Going back down, the net inward pressure resealed the gasket, so air couldn't reenter. We may have had as much as 5 pounds per square inch holding the lid on, which would have come to over *five hundred pounds* for the lid as a whole! The cure, by the way, was simple. Customs, with fine forbearance, allowed me to take the trunk to the lab, where a long and satisfying hiss of inhalation followed my drilling a small hole. The top then allowed itself to be lifted off without protest.

Most of the time, however, we can ignore the background pressure of the atmosphere. No terrestrial organism is so tall so that the air pressure on upper and lower extremities is appreciably different (although one can't make the same categorical statement for aquatic creatures.) Birds, though, with a penchant for swooping and diving, do have provisions for venting the enclosed air in their hollow bones. We ignore atmospheric pressure so consistently that values of pressure commonly quoted assume as a preexisting pressure that of the atmosphere. The 26 or so pounds per square inch in your tires (*gauge pressure*) is entirely above the 14.7 PSI base line; your blood pressure of, say, 120 mm Hg presumes a preexisting pressure of around 760 mm Hg. Using atmospheric pressure as a baseline is matter of both practicality and candor. Practicality, because the basic atmosphere (and its variations) is practically irrelevant in most situations; candor because pressure gauges such as U-tube manometers usually measure differences in pressure rather than absolute pressures. Most often you don't measure pressure per se but rather the difference between some pressure of concern and that of the surroundings. Thus the free end of the U-tube of a sphygmomanometer is open to the atmosphere.

Another unit of pressure while less common (although in use by fluid mechanists) is nicely revealing of how and why certain changes in pressure occur. Let's fill that U-tube, not with mercury again, but with water, which is 13.6 times less dense than mercury. My blood pressure would cause the water columns to differ in height not by 110 millimeters (a little over 4 inches) but by 13.6 times as much—about 1.5 meters or almost 5 feet. That's a little awkward, which is why mercury is in use. (Mercury doesn't evaporate rapidly, which is another advantage.) A full atmosphere of pressure would raise the column by 760 mm times 13.6—10.3 meters or 33.9 feet. SCUBA divers will recognize this as a familiar datum. If you descend 33.9 feet beneath the surface of a body of water, then the pressure on you (all over—pressure is omnidirectional) increases by an atmosphere. If the dive was begun at sea level, then the pressure on you has doubled to a total of two atmospheres. The pressure increases by another atmosphere for each 33.9 feet of descent.[5] The point is that depth is depth, and pressures don't know the difference between lakes and manometer columns.

Now the people who take my blood pressure have been taught to keep the part of my arm with the cuff just opposite my heart, neither higher nor lower. They know that to do otherwise imparts some error in the measurement, but few of them seem to have much notion of the magnitude of the error. If I've had a salty dinner the night before, I slouch down in the chair, so my arm on the table is a little higher. Blood has about the density of water, and I'm just your ordinary middle-aged, ambulatory manometer. If the cuff is 5.4 inches higher than my heart, the pressures are 5.4 inches or 136 mm of water lower at the cuff than at my heart. This 136 millimeters of water (divide by 13.6, the ratio of the density of mercury to that of water) corresponds to 10 mm Hg. Not to complicate the point, I contrive to reduce the readings by 10—from, say, 120 over 80 to 110 over 70—by raising my arm a little over 5 inches.

Having gotten this far into the practical matter of cheating at manometry, it's worth mentioning just how the cuff-and-stethoscope trick gives blood pressure. What's of interest, of course, is pressure just downstream from the heart—in the aorta. The major arteries have little resistance to flow, so pressure doesn't change appreciably as blood moves from the aorta to the large artery of one's upper arm. To start with, the cuff is inflated to a pressure in excess of that at which the heart is pumping. The pressure is transmitted through the wall of the cuff, your skin, and your muscle (none of which have much effect on it) and acts on the artery carrying blood from the heart (as well as on everything else in the vicinity, as you can feel). As a result, the artery collapses, since pressure inside is now lower than pressure outside. The pressure in the cuff is then slowly reduced while someone (even oneself) listens, immediately downstream from the cuff. When the pressure in the cuff falls just below the peak pressure delivered by the heart as it contracts, blood begins to squirt through the obstruction. The turbulence of squirting makes noise, which can be heard. As the pressure falls further, the sound continues; but eventually it ceases. Cessation of sound means the end of turbulent flow and signals full, normal flow through an artery that is not appreciably constricted any longer. And that means that even the minimum pressure of blood coming from the heart—when the ventricles are refilling—is sufficient to get past the cuff. The points when the sound begins and ends are reported, thus the maximum and minimum pressures in the great aorta just downstream of the left ventricle. These extremes of pressure are called the *systolic* (maximum) and *diastolic* (minimum) values.

Getting the extremes of pressure at the heart with neither fancy gear nor invasive procedure is really quite a neat trick. It takes advantage of some of the peculiarities of pressure that have just been described, in particular that pressure is exerted equally in all directions. It also manages (if done properly) to circumvent any serious gravitational effect.

But Pressure Poses Peculiar Problems

Still, gravity is ever present, and sometimes its effects are, one is tempted to say, matters of real gravity. Our heads are normally further from the center of the earth than our hearts, so arterial pressure at the brain is lower by the expected amount—around 30 mm Hg or 16 inches of water. The difference is occasionally of significance. At times my own blood pressure decides to go through a low phase; when that happens I find that I have to get out of bed with deliberation or I'm dizzy when I first stand. It seems to take a little time for the system to readjust and to generate enough pressure for the brain to direct navigation safely.

In some animals the vertical heart-to-head distance is still greater than ours, most notably, of course, in giraffes. The founder of modern comparative physiology, August Krogh, of Denmark, wondered about the cardiovascular dynamics of giraffes in the 1920s, but he wasn't in a position to do any direct measurements. Further interest came during the Second World War, stimulated by a practical problems of pilots. In situations in which planes pulled out of dives the force of gravity effectively increased. Increasing the value of gravitational attraction, as every science fiction fan knows, is equivalent to increasing your height. Such an increase in gravity, as you can by now guess, causes pressure to decrease more than it normally would as blood ascends from heart to brain, so healthy young pilots with nicely low blood pressures (= "nonhypertensive") showed awkward tendencies to pass out at critical moments.

For giraffes, of course, drastic decrease in pressure between heart and head should be an everyday matter, something for which appropriate compensatory machinery must have long ago evolved. In the 1950's and 1960's people began to do actual measurements, including telemetry of data on pressure and flow from unrestrained animals. Adult giraffes stand 15–18 feet tall, with the heart about halfway up. They are, it turns out, normally hypertensive, at least by ordinary mammalian standards. Their systolic pressures at rest are about 200–300 mm Hg, and their diastolic between 100 and 170 mm Hg, both about 100 or so millimeters higher than ours and sufficient to pump blood up an additional 5 or 6 feet.

The giraffe's brain is thus supplied with blood at about the same pressure as our own. Still, that's only one end of the problem, and the giraffe must contend with the problem of high pressures lower down. (You can easily calculate the pressure in an artery just above the hoof.) In practice, the giraffe responds by building stronger walls on its pipes and by encasing its extremities in a strong and minimally extensible skin. Extremities include head as well as feet, since the animal has to lower its head to drink without (as my friend Mark Denny puts it) blowing its brains out. The giraffe has invented something analogous to what military aviation calls a "g-suit"—a tight garment that offsets excessive gravitational forces. Squeezing inward on torso, legs, and arms provides extra pressure to push blood up to the head. Support stockings, by the way, are a mild version of the same thing. A biologist could puff up a bit if development of the g-suit traced to research on giraffes; in truth, the suit preceded the work on giraffes. In fact, that's been the usual sequence, with biologists discovering natural analogs of existing bits of technology but failing to recognize devices that happen to lack contemporary technological examples.

Not unexpectedly, giraffes are just the extreme of a continuum. For instance, horses have higher systolic blood pressures than humans (I recall seeing a value of 183 mm Hg quoted), and elephants wear strong support stockings. Giraffes, however, are far from the longest mammals, a record clearly held by whales, which can reach 100 feet. Moreover whales must extend that length vertically in rapid diving and ascent. A whale, on the other hand, hasn't the giraffe's problem since it lives in water, not air, and is essentially aquatic both inside and out. In any posture, gravity has the same effect on the water outside a whale as on the blood within it. Diving by air-breathers raises all sorts of problems connected with compression of air in the lungs, with dissolution and bubbling out of nitrogen in the blood, and with the simple inability to obtain oxygen beneath the surface. Giraffes and g-suits, though, are irrelevant.

Another case (really a set of cases) is in its way even queerer. Reptiles have lower blood pressures than mammals. Dismissing low pressure as merely primitive is not entirely fair. As noted earlier, reptiles don't spend vast amounts of energy maintaining a high and constant body temperature and can do with a less highly supercharged circulatory system. (If reptiles were inferior in some absolute sense, they surely would have gone extinct in this evolutionarily competitive world!) Most reptiles either live immersed in water or stay more or less horizontal, so local variations in blood pressure should be minor. Still, a few elongate ones—snakes—on occasion engage in such unreptilian activities as climbing trees. Problem—how can a snake climb a tree without passing out? It turns out that both their structural arrangements and their functioning have changed in fairly consistent ways on each occasion that a lineage of snakes has taken on arboreal habits.

Snakes generally maintain a blood pressure in the arteries entering their brains of about 30 mm Hg—not much different from ours. Aquatic and semi-aquatic snakes do so with hearts located about halfway back from head to tip of tail. In both terrestrial and, more drastically, in arboreal species, however, the heart is farther forward, getting as far anterior as a seventh of the way from head to tail. This dramatic adjustment in anatomy means that arboreal snakes don't need to generate especially high pressures at their hearts to keep their brains properly supplied. Even so, their systolic blood pressures are about twice that of the fully aquatic sea snakes—in water of course, there is that neatly counterbalancing hydrostatic pressure, so even if a sea snake swims directly upward, it encounters no gravity-related pressure shifts. Furthermore, the systolic blood pressure of climbing snakes depends on body length. The longer ones have higher pressures, whether one considers differences in pressure between species of different adult lengths or the changes in pressure as an individual grows. That's just what you'd expect, since a long snake has (if constructed in the same proportions) a head further from its heart than does a short one. And, finally, the climbing snakes have a relatively inextensible body wall and skin—support stockings (stocking?) again.

In both mammals and snakes, blood pressure is the object of active regulation. Receptors ("baroreceptors") in various locations sense pressure levels, and the relative resistance to flow of different parts of the body are adjusted to restore normal pressures after some disturbance. In air (but not when weightless in water or spacecraft) change in posture with respect to the earth can cause major local pressure changes. One can impose such a change in an especially uncomplicated fashion by tilting a restrained snake. A head-up tilt gives an immediate reduction in arterial blood pressure, but that's followed by rapid adjustment so pressure goes above the normal value—enough to give a proper supply to the brain. At least that's what happens in an arboreal or semi-arboreal species. By contrast, a sea snake tilted head up in air is in a bad way; it doesn't give the proper response, and the blood pressure in its head may even fall below the local atmospheric level. This may have relevance to a safe technique for handling sea snakes, if you must do so, which I most emphatically don't recommend.

Lots of data are available on blood pressures among animals. Those of birds run generally higher than mammalian pressures, but owls, for some reason, have lowish pressures. Fishes are a diverse lot, so it's hard to generalize. Most arthropods, even large ones, have low pressures (a lobster gets up to only 11 mm Hg or thereabouts); and mollusks are mostly very low, generally less than 3 mm Hg. Still, a few invertebrates do go in for well-pressurized circulations. An exercising octopus can generate systolic pressures of around 80 mm Hg. Some extraordinary earthworms living in tropical South America reach fully 4 feet in length and over 1 pound in weight. When active these generate blood pressures of over 90 mm Hg. The most extreme cases are spiders, although here one is talking about internal pressure everywhere, since they mostly lack discrete return pipes in their circulatory systems. Spiders, it turns out, use no muscles to extend their legs, sticking them out hydraulically instead. The pressures involved sometimes exceed 450 mm Hg, far higher even than the systolic pressure of a giraffe.

Finally, Where Does the Pressure Go?

Consider a normal human or other ordinary mammal with systolic and diastolic blood pressures of 120 and 80 mm Hg, respectively. Clearly these aren't the pressures throughout the circulatory system, for if they were, no blood would feel pressed to make the circuit. They are, to remind ourselves, the pressure extremes in the aorta and the other large systemic arteries. How, then, does pressure vary, not just within the period of a heartbeat, but in the various parts of the system? First, the systolic pressure does indeed reflect the maximum pressure in the left ventricle, here 120 mm Hg. During diastole, by contrast, the pressure in the left ventricle is not far from zero; so the aortic valve is held shut by a pressure difference of fully 80 mm Hg. (Assuming a valve area of around a square inch, that's a force of about 1.5 pounds.)

Beyond the large arteries the pressures drop sharply, and the diastolic and systolic pressures gradually converge. By the time blood reaches the capillaries, the fluctuations are minor, and the pressure has gone down to about 25 mm Hg. A lot of variation hides behind that number; nevertheless it's clear that the largest part of the pressure drop in the entire circuit is in the small arteries, the so-called arterioles. Pressure drops a little more in the capillaries, down to very roughly 15–17 mm at their venous ends, although it ought to be noted that the pressure drop per unit length of pipe is especially great in these relatively short vessels. (The notion of pressure decrease per unit length will take on more significance in the next chapter.) Most of the rest of the pressure drop occurs in the smallest veins (*venules*) with very little further drop in the big veins. By the time blood gets back to the vena cavae, pressures are around 4 mm Hg. That's not a lot, but the right atrium has a bit of resilience—like a baster bulb it prefers, because of wall elasticity, not to be collapsed—so its internal pressure can fall below zero (actually, of course, merely below atmospheric pressure, the conventional base line.) Thus a bit more than a 4-mm pressure difference is available to fill the atrium.

Blood pressures on the pulmonary loop are much lower, even though the overall volume flow rates have to be the same. It's a circuit of lower resistance and uses a weaker pump—recall that the right ventricle has a wall only about a fourth as thick as that of the left ventricle. Peak pressure in the right ventricle or a pulmonary artery is about 20 instead of 120 mm Hg. Again the pressure in the artery never drops as low as that in the ventricle, so the average pressure of the blood supply to the lungs is about 15 mm Hg. In the pulmonary veins, the pressure is about half of that average, which means that the left atrium gets blood at a somewhat higher pressure than does the right one. That may be of importance since of course the left heart has to boost pressure to its highest value anywhere.

All of these pressures have to be adjusted for gravitational effects, of particular importance in upright creatures such as ourselves. The adjustment is now familiar: 10 mm Hg drop for each 13.6 cm or 5.4 inches of elevation. The adjustment, if you try it, reveals a peculiar thing. In the capillaries of both the head and the upper parts of the lungs, pressure is normally subatmospheric! Thus pressure is greater outside than inside, giving a net collapsing force. As with the right atrium, sufficient vessel resilience is apparently present to counteract this alarming possibility.

* * *

One can tell other tales of pressure and circulatory systems, but most involve the operation of bits and pieces of the machinery that we have yet to introduce. Instead, we'll move on to make pressure do that for which hearts bother to pump it up in the first place—to force fluid, blood in particular, to flow.

Notes

1. Chemistry is irrelevant. Soap doesn't cause slip of the present sort, affecting only the interaction of solid surface, liquid, and air. With some dissolved detergent, water won't "bead" on the drained surface and various deposits will themselves dissolve, but the basic flow is unaffected.

2. But windowpanes, like most of us, get thicker with time at their lower ends.

3. Solids, liquids, gases—all three states of matter are relevant to the present subject. A solid has both shape and size; a liquid has size but no shape; a gas has neither size nor shape. Liquids and gases are collectively called *fluids*, a useful lumping since both will flow and have viscosity, distinguishing them from solids.

4. The *newton* (after Sir Isaac) is the accepted unit of force for scientific work. It corresponds to the downward force of a 102-gram mass (about 3.5 ounces) at the surface of the earth—roughly a quarter of a pound.

5. Mention of the sea reminds me to note that the dive above was in fresh water; sea water is denser—10.1 meters or 33.1 feet correspond to an atmosphere of pressure.

5 *How Blood Moves*

Fluid mechanics is the branch of physical science that assures us that airplanes can fly, that flags must flap, and that you'll fall no faster after a 1-mile drop than after just a few hundred feet. The subject treats both gases and liquids and thus the two fluids, air and water, that are of special concern to biologists. In particular, it's the source of the rules that constrain the ways blood can flow.

Despite its everyday applicability, fluid mechanics turns out to be one of the stranger areas of physics, replete with wondrous and counterintuitive phenomena. Most of us regard physical science as perhaps abstruse but certainly tidy—it's a bit boggling to find that ordinary aspects of something as ordinary as water flow are not at all easily brought under the lawful command of straightforward equations. You may know the Newtonian laws of motion and gravitation, you may be comfortable with the conservation laws for mass, momentum, and energy, and you may be able to make vectors veer in practically any direction. Yet none of that will dependably enable you to predict the pattern of flow over something as simple as a sphere. Still, we do build airplanes to which we trust our nearest, dearest, and even ourselves; and our sailboats outperform those of the cleverest designers of antiquity. Moreover, the very peculiarity of the subject gives nature rich opportunities to surprise us, for organisms to show that they're indeed those cleverly adapted little rascals for whom we biologists have such great affection.

Here, then, is a chapter about fluid mechanics, at least the parts of it central to the subject of circulation. In fact, we've already introduced two of the relevant quantities, pressure and viscosity. The only other property that matters much from the present vantage point is density, and that was alluded to in connection with manometry. *Density*, to remind ourselves, is simply the mass of a sample of some material divided by its volume. A cubic foot of water weighs a little over 62 pounds; a cubic foot of mercury weighs 13.6 times as much. Their densities are thus 62.4 and 816 pounds per cubic foot or 1000 and 13,600 kilograms per cubic meter.

Fluids in Motion Don't Shrink or Swell

Fluids, again, are oddly behaved accretions of matter. We've already run into a truly nonobvious concept, the no-slip condition. Another is needed at this point, one that very nicely simplifies the way that fluid mechanics applies to organisms. Most readers will be familiar with a bicycle hand pump. Push on the handle and air squirts out into a tire. If you put your thumb over the orifice and push on the handle, the piston still goes down, although not so far and with greater difficulty. You are, as it happens, compressing the air in the cylinder—a given mass of air is being made to occupy a smaller space, or, to put it another way, the air's density is increased by your push. Clearly air is compressible stuff. Water is much less so.

The assertion we make is that for problems involving the flow of fluids, *we can get away with treating all fluids, including both air and water, as if they were incompressible.* The assertion can be shown to be reasonable at least for problems where the speed of the motion is low relative to the speed of sound. Since sound travels far faster in either air or water than any creature swims, flies, or pumps its blood, we're on very safe ground. The main consequence of the fact that fluids suffer no significant compression by these slow flows is that their densities are constant. In short, whether pushed by elevated pressure upstream, pulled by lowered pressure downstream, or accelerating or decelerating, 62.4 pounds of water will still occupy a cubic foot.

The resulting simplification is this. In the larger world of physical and chemical processes, mass is conserved, which is to say that the same mass is present after some treatment as was present before. (I'm ignoring nuclear reactions as being of no household relevance.) If you burn something and add up the mass of gases released and ash remaining, the sum is the same as the mass of fuel and oxygen consumed. *For our sorts of problems of moving fluids, conservation of mass implies conservation of volume as well.* That's the implication of incompressibility. A gallon of blood has the same mass whether at the high systolic pressure of the left ventricle or at the slightly subambient pressure of a relaxed right atrium; in the sense of *either* volume or mass, it is the same amount of blood.

We can take the matter a step further. If you have a pipe connecting two bodies of fluid—perhaps a straw through which you sip a drink—the mass of fluid entering one end of the pipe must be the same as that leaving the other over any period of time you might care to consider. Incompressibility means that the volume of fluid entering must be the same as that leaving also (assuming, of course, that the straw doesn't swell or shrink). Volume per unit time going in must equal volume per unit time going out. Too, a step still further, the volume per time, the volume flow rate, is the same at every place along the pipe. That is, if the pipe were to be cut and a flow monitor inserted, the same volume flow rate would be measured at any cross section. If 1 gallon per minute passes one location it must pass any other as well. This trivial-sounding notion proves so important in fluid mechanics that it's given its own name, the *principle of continuity*.

Actually, we invoked the principle earlier. It demands that in a circulatory system that has its components serially arranged, the volume flow rate must be the same in each component of the series. Flow goes from right atrium to right ventricle to lungs to left atrium to left ventricle to body capillaries and back to right atrium, and no component can be bypassed. The flow rate must therefore be the same in each. If the left ventricle pumps 1 gallon per minute, so must the right, and the flow through the lungs must also be 1 gallon per minute as must the overall volume flow through the rest of the body's capillaries. You can now see why the lungs, a tiny fraction of the weight of the body, get their blood through a pipe, the pulmonary trunk, of an internal diameter just about the same as that of the great aorta that supplies everything else—one of the awkward anatomical facts that Harvey faced up to properly. It's obvious that some bypassing or shunting connection is absolutely necessary if a mammalian fetus is to avoid having to pass such an enormous amount of blood through its nonfunctional lungs.

As adults, birds and mammals may have different pressures in their systemic and pulmonary circuits, but they must have the same volume flow rate. We're not the only vertebrates that use lungs, though, but while many adult amphibians and all reptiles are lung-using air breathers, none have our fully serial, nonshunted circulations. Systemic and pulmonary circuits are to one extent or another connected. The connection may be through the use of a single ventricle, as in amphibians, through an incompletely partitioned ventricle, as in most reptiles, or just through a cross connection beyond the ventricles as in the crocodilian reptiles—Figure 5.1 gives a few of the particulars. With the anthropocentricism to which even biologists sometimes succumb, these arrangements are frequently viewed as primitive, which is certainly unjustified. Even worse, although not commonly done any more, they have been interpreted as anticipating the avian and mammalian ideal, which is really nonsense—natural selection is simply and totally incapable of foresight.

Once again we're certainly looking at different ways of making a living, the difference between the hot and frenzied birds and mammals and the cool and patient reptiles and amphibians. It's quite likely that we've lost something by adopting our unalterably serial circulation. A diving mammal cannot shut down its lungs when under water without shutting down the rest of its circulation to the same extent. A diving iguana can make such adjustments,[1] and a frog underwater can adjust its circulation to make use of its skin as a gill, while largely bypassing its quite useless lungs. At the same time, however, we've also gained something. With no interconnection of pulmonary and systemic circuits, the pressures in the two can be quite different. In short, pressure or volume flow rate can be different in the two circuits, but not both—one or the other quantity must be constant. Reptiles and amphibians have opted for different flows and similar pressures. Birds and mammals have chosen instead to have different pressures and to put up with the limitations of equal volume flow rates. It's possible that our use of high systemic blood pressure would, if carried into the pulmonary circuit, impose some unmanageable burden or risk.

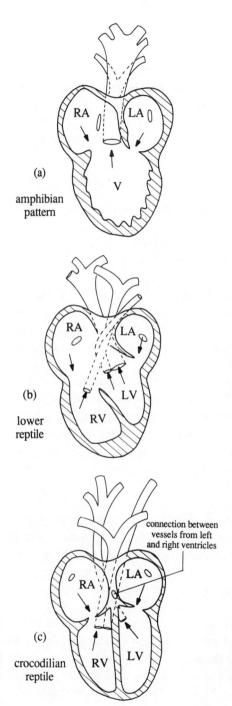

Figure 5.1. Blood may move in a simple serial circuit in fish, birds, and mammals, but the other vertebrates aren't quite so simply set up. Here are the hearts of (a) amphibians such as frogs, (b) most reptiles such as snakes and turtles, and (c) the crocodilian reptiles in particular. While in each case oxygenated (body-bound) and deoxygenated (lung-bound) blood isn't completely separated, the extent of the interconnection varies greatly. Amphibians have a single ventricle, the "lower" reptiles have a partial partition of their ventricle, and the crocodilians have only a small opening interconnecting pipes from left and right ventricles as they leave those chambers. This latter runs front to back and is indicated as an oval in the wall between the two sides of the heart. (Abbreviations: RA = right atrium, LA = left atrium, V = ventricle, RV = right ventricle, LV = left ventricle.)

One lesson from this comparison is that *primitive* and *advanced* are dangerous terms. They're certainly not automatically synonymous with *ancestral* and *derived*, which are irreproachably historical and bear no judgmental hubris. A former student, Terri Williams, suggests that we use *obstinate* and *capricious*—in effect proclaiming a plague on both your houses. I rather like that.

Flow in Nozzles and Branches

According to the principle of continuity, volume flow rate is the same everywhere in a serially arranged sequence of pipes. Obviously if a pipe has a fat region, there's more pipe through which fluid can flow. A lower speed of flow is therefore adequate to get the same volume flow rate. Conversely, in a narrow region the speed of flow must be greater, as shown in Figure 5.2a. That's how a nozzle works—when you want faster flow in order for the stream to squirt further you just constrict the channel. It works for garden hoses, shower heads, water pistols, aerosol cans, air jets, hair dryers, leaf blowers, and so forth. If your teapot has a spout that gradually tapers down toward its orifice, it will more likely pour a satisfactorily fast and coherent stream. This works wherever an organism squirts—the water-ejecting jet of a squid, the anal jet of a dragonfly larva, the squirt of water shot into the air by an archer fish. The operative rule is simple. The result of multiplying speed of flow times the cross-sectional area of the pipe must remain constant, since that result is the volume flow rate.[2] Make the cross-sectional area smaller, and the speed of flow must increase in the same proportion.

You may protest that the capillaries are *really* small, so the principle would have us expect horrendously high speeds of flow in them. Does that actually happen? It doesn't, because of another piece of the same story. Everything previously has referred to serial arrangements. Capillaries, by contrast, are parallel elements—a bit of blood going out of the aorta might pass through any of a couple of billion capillaries before emerging into a vein. So what matters to flow speed isn't the cross section of a single capillary, but the combined, aggregate cross-sectional area of all of the capillaries open to flow at a particular time—the situation is similar to that shown in Figure 5.2b. Since the combined area

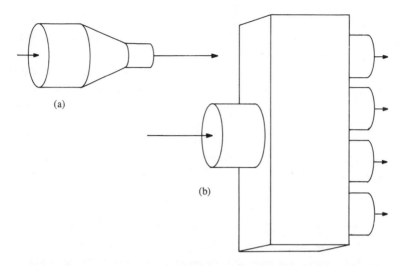

Figure 5.2. (a) Increasing flow speed through the use of a constriction in a pipe, the nozzle notion. (b) Decreasing the flow speed despite pipe constrictions by using enough daughter pipes so their total cross-sectional area is greater than that of the parental pipe. The length of each arrow is proportional to the local speed of flow.

is very large, the flow speeds in capillaries are quite low. These low speeds affect both the time available for material to pass between capillaries and tissue and the cost of pumping blood through the capillaries. We'll return to both of these important subjects shortly.

This principle, that the product of cross-sectional area and speed must be the same at any level in the circuit, permits a most interesting crude calculation. Say we measure the cross-sectional area of the aorta, which is easy, and the speed in the aorta, which is at least possible. Or, much the same thing, we determine the volume flow rate of the heart from stroke volume and heartbeat rate. We also determine the speed of flow in a typical capillary, which is not especially difficult. We can then calculate the aggregate cross-sectional area of the body's capillaries. By taking the aortic area as 5 square centimeters and the average speed of flow there as 20 centimeters per second we get an overall flow rate of 100 cubic centimeters per second (about 1.5 gallons per minute). If the flow speed in a capillary is half a millimeter per second (an inch every

50 seconds), then the total capillary cross-sectional area must be about 2000 square centimeters, or a little over 2 square feet. Putting the matter another way, flow has slowed 400-fold, so area must have increased just 400 times. Finally, the cross-sectional area of a single capillary is well known, and we can easily divide that 2000 square centimeters by the area of a single capillary. What we get is an estimate of the number of capillaries in the body. Not to keep anyone guessing, the number comes out to about 7 billion. It may be a rough estimate, since it ignores details such as arteriovenous shunts and so forth, but calculating it using the principle of continuity surely beats counting.

The Character of Flow

Fluid flow, again, is a most curious branch of physics. We turn now to quite another peculiarity, which is that how a fluid flows depends on the scale of operation of the flow. By "how it flows" I mean not just a quantitative measure such as speed, but instead the entire qualitative character of the motion. For flow in pipes two fundamentally different kinds of flow occur, one termed *laminar,* the other *turbulent.* The difference is perhaps easiest to envision by referring to the classic experiment of Osborne Reynolds,[3] reported in 1883. Figure 5.3 illustrates the centerpiece of Reynolds' set-up. In both conduits a liquid is flowing through a pipe from left to right, and in both a stream of dye is steadily injected through a fine tube into the flow. At the top the dye moves down the middle of the pipe with little spreading of the dye stream; at the bottom the dye stream almost immediately breaks up and colors the entire flow. Above, then, flow is laminar; below, it is turbulent. Above, every bit of fluid moves in the same direction (although not necessarily at the same speed); below, we can recognize an overall direction, but that's only a kind of average, with motion in any direction possible locally and briefly. Above, flow is not an intrinsically disordering process; below, mixing is its inescapable concomitant.

Naturally the big question is what determines which kind of flow occurs in a given situation. From your everyday experience you might guess that high values of viscosity predispose a fluid toward laminar flow—years of experience mixing cake batter can't be denied. Indeed,

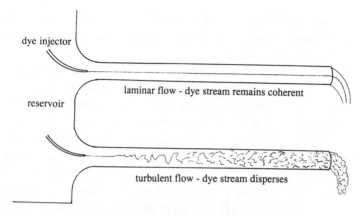

dye injector

reservoir

laminar flow - dye stream remains coherent

turbulent flow - dye stream disperses

Figure 5.3. Two regimes of flow through a pipe, laminar and turbulent.

Osborne Reynolds discovered about 100 years ago that higher viscosity in a fluid opposes turbulence, whereas higher density makes turbulence more likely; however, the regime of flow isn't determined solely by the value of viscosity. In particular, and here's where the scale of operations takes on importance, large size and high speed both favor turbulent flow. From his measurements Reynolds derived a variable, now called with decent respect the *Reynolds number,* whose value is a good index for telling whether flow will be laminar or turbulent. The variable isn't complex—it's simply the density of the fluid times the diameter of the pipe times the speed of flow divided by the viscosity of the fluid. At high Reynolds numbers flows are turbulent, while at low ones they're laminar.

In a physical sense, what determines the transition from laminar to turbulent flow is the balance between two conflicting tendencies in the fluid. On one hand, fluids resist deforming motion, which is to say that little bits of fluid like to move together; it's this "groupiness" that viscosity's about. On the other hand, the bits of fluid like to continue moving as they have been, whatever any other bits are doing. This "individuality" is what Newton's first law of motion speaks about—a mass continues to move in a straight path unless acted on by an external force. The Reynolds number boils down to the ratio of these two tendencies that I've loosely called (at the suggestion of a former student, Kate Loudon) individuality and groupiness. At high numbers, the first predominates, with just the sort of anarchy a social scientist would anticipate. By contrast, at low Reynolds numbers the second holds sway, and flow's a proper parade.

(Incidentally, flows of air and water aren't dissimilar. Water may be denser than air, but it's more viscous as well. Thus the ratio, what matters, isn't so terribly different for the two media when they're flowing through pipes of the same size at the same speed.)

This qualitative difference in regimes of flow can bedevil one's view of what must happen in circulatory systems. We're very big creatures, as organisms go; familiar aspects of our culture and technology are similarly large. The flows most familiar to us are turbulent—in bathtub and vacuum cleaner, around car bodies and boat hulls—where flow is intrinsically messy. In our size range one only occasionally encounters laminar flow, such as when pouring a liquid of high viscosity, what we normally call a syrup. The very orderliness of such flow may be a positive nuisance—try to mix dark molasses and light corn syrup with both at refrigerator temperature. Conversely most of what happens in circulatory systems happens in a laminar regime, and the discussion to come presumes laminar flow—except for the principle of continuity, the rules for turbulent flow are different.

Things may get a bit turbulent during vigorous exercise or for very large mammals, but even the flow through our largest pipes is normally laminar. We're close to the transition point, though, and in some not uncommon pathological conditions turbulence does occur in the aorta. Turbulence is noisy, as noted when we considered how blood pressure was measured. The result, in addition to noise, is a bit of doggerel quoted by Alan Burton, from the old days when medical students were presumed male:

> Streamline flow is silent;
> Remember that my boys!
> But when the flow is turbulent
> There's bound to be a noise.
>
> So when your stethoscope picks up
> A bruit, murmur, sigh;
> Remember that it's turbulence,
> And you must figure why.

Let's return to the relative arrangements of the hearts of the various air-breathing vertebrates, looking again at Figure 5.1 and bear in mind that laminar flow is not necessarily an effective mixing process. Having two fluids flowing in proximity without mixing may seem strange, but at relatively low Reynolds numbers it's apparently not hard to accomplish. Indeed, the right atrium of a mammalian fetus does the trick, as mentioned in Chapter 2. Now the old accounts of the circulation of the amphibians and reptiles inevitably talked about mixing of pulmonary and systemic blood in the heart, and the inefficiency of combining oxygenated and deoxygenated blood. Sending partially oxygenated blood off to both lungs and elsewhere certainly sounds like bad design. But no evidence for the mixing seems to have existed; it was just what our intuition assured us would happen. Again, we're immersed in an external world of high Reynolds numbers, the world of winds and streams and car-washes and hair-dryers, and we've got the wrong intuitive picture. The expected mixing just doesn't take place, at least to any great extent, according to some recent and fairly incontrovertible measurements. So regarding the arrangement as inefficient is no longer defensible, bolstering the view taken earlier that we're just looking at a different functional scheme, appropriate for a different style of living than that of birds and mammals.

Even before real turbulence sets in, higher Reynolds numbers imply a greater mixing of adjacent elements of a moving fluid. This purely physical phenomenon may explain some further differences in the circulatory arrangements of these various vertebrates. As one moves from amphibians to crocodilians, partitioning of the two circulatory circuits increases, from the completely undivided single ventricle of a frog to the small opening between paired aortas in an alligator. Perhaps what we're seeing is mainly an effect of scale, a consequence of differences in Reynolds number rather than of accidents of ancestry. On average, amphibians are both smaller and more sluggish than are reptiles, so their hearts are smaller and blood flows more slowly. Mixing is therefore less likely, and more contact between circuits is tolerable. (Still, the arrangement to minimize mixing in frogs is fairly complex, as shown in Figure 5.4). Crocodilians include the largest of extant reptiles. Mixing is therefore more likely, and the heart is arranged to minimize it, while still providing some provision for shunting blood from one circuit to the other. It would be interesting to know whether there's more mixing in the hearts of the larger members of any fairly homogeneous group of reptiles or amphibians—or, conversely, whether the hearts of the larger species have relatively smaller shunting connections. Biology can't transcend physics, but the operation of physics sometimes may be superbly subtle.

Pressure, Flow, and Cost

We now have enough pieces assembled to take a look at a slightly more elaborate principle, one that interrelates the pressure pushing fluid through pipes, the amount of flow resulting from such a push, and the cost of giving the push and getting the flow. The basic idea (which we can now easily calculate from theoretical considerations) came from very careful experiments done about 1840 by Hagen, a German, and Poiseuille, a Frenchman. Neither was immediately aware of the work of the other; since they got the same results Europe wasn't noticeably more turbulent. Unfortunately the equation that we now use is named either Poiseuille's equation or the Hagen–Poiseuille equation. "Hagen" gives no trouble, but "Poiseuille" can be correctly pronounced only by birthright natives, and both HP and PH as acronyms have been preempted.[4]

The particular equation works only for laminar flow. In the strictest sense it also assumes that the walls of the pipe are rigid, that the flow is steady, that no entrance or branch point is nearby, and that the fluid is homogeneous and has no trace of the elasticity of solids. Laminar flow is pretty much okay, but all the other conditions are to some extent violated in circulatory systems. That's the usual way when we apply equations to organisms—the equations apply to situations that aren't exactly what really goes on. In practice a generous dose of skepticism and circumspection has to be employed antidotally to the resplendent elegance of mathematical modeling. It's yet another reason why we have to resort to direct experimentation and measurement on

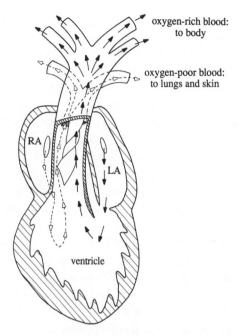

Figure 5.4. A contemporary view of how two sorts of blood flow simultaneously but largely unmixed through the ventricle and aortas of a frog. Ninety percent of the output of the left atrium goes to the lungs.

living organisms as adjuncts to even the most brilliant of simulations using the most powerful of computers.

Perhaps the easiest way to put the business is to ask what determines the volume flow rate through a pipe. First, it turns out that the rate is proportional to[5] the pressure difference between the ends of the pipe divided by the length of the pipe—the pressure drop per unit length. Other things being equal, if a heart were to double the pressure difference between left ventricle and right atrium, then twice as much blood would flow through the systemic circuit. (In practice, we do increase pressure to generate more flow, but it's not the only change, nor even the most important one.) The greatest pressure drop per unit length of the systemic circuit is in the capillaries—twice that in arterioles and venules and at least an order of magnitude larger than that in the larger vessels. But that doesn't mean that by far the greatest part of the overall pressure drop per se occurs there. The practical way to compensate, which we do, is to keep capillaries short; a typical one is only about 1 millimeter long.

Second, the volume flow rate is inversely proportional to the viscosity of the fluid. Thus if blood viscosity were doubled, then flow would be half as rapid. As it is, our not-atypical blood is already somewhat more viscous than pure water, by a factor of 3 or 4—a datum recognized, most notably, by Sir Walter Scott. An interesting issue arises in connection with blood viscosity. As mentioned, the main function, quantitatively speaking, of blood in animals anything like ourselves is carrying oxygen from lungs elsewhere. Oxygen is carried attached to hemoglobin (the red stuff), and hemoglobin is entirely contained in the circulating red blood cells. About 45 percent of our blood is made up of red blood cells (that percentage, to remind you, is what's meant by the term *hematocrit*). With more red blood cells, more oxygen can be carried. Or, to put the matter pathologically, if you lack sufficient red blood cells, you can't sustain normal levels of activity. Why not put more cells in the blood? The answer seems to be that the cost gets disproportionate to the benefits—the viscosity goes up at an ever increasing rate when the hematocrit is raised further, and much more pressure is necessary to get the same volume flow rate. Diving animals such as seals and whales do have a somewhat higher concentration of red blood cells than do we. In their situations, though, the ability of hemoglobin to act as a short-term oxygen store is especially crucial, and somewhat greater viscosity might reasonably be worth the extra cost of pumping. Breath holding is the name of the game if you persist in living much of your life underwater but still insist on breathing atmospheric air.

Finally, the volume flow rate varies with the diameter of the pipe. Here we run into something far more drastic than ordinary proportionality. Assuming the same pressure drop per unit length and the same viscosity of fluid, a pipe of twice the diameter of another will carry not twice, nor four times, nor eight times, but fully *sixteen times* the flow. A pipe increased in diameter threefold will carry $3 \times 3 \times 3 \times 3$ or 81 times as much fluid per unit time. Put formally, volume flow rate is proportional to the fourth power of the diameter of a pipe. Or, putting it the other way, small pipes are in serious trouble; a designer ought not use small pipes unless flow is very slow or some other consideration is of critical importance. Despite this seemingly practical prohibition on small pipes, organisms, whether trees, sponges, squid, or people, are liberally endowed with exceedingly tiny pipes. To explore the rationale and ramifications of systems of small pipes some further physical notions are needed. So I'll defer the matter to the next chapter, where it will get us into the very core of the rationale for having fluid transport systems at all.

I should mention that this fourth-power rule isn't quite as directly applicable as it's sometimes made out to be. With serial sequences of pipes, increasing the diameter of one pipe may not increase the flow rate much at all. That happens when the other elements of the system provide the main resistance to flow. Replacing your garden hose with fire hose will have little effect on how much water is delivered. The main items of resistance are certainly elsewhere, since the house probably has a long run of water pipes of about a half-inch internal diameter and since both faucet and nozzle involve very much smaller orifices. Even if your coronary arteries are 50 percent occluded with fatty deposits, you may still feel no special disability, certainly nothing that would indicate any 16-fold reduction in blood flow—the bulk of the resistance to flow resides elsewhere in their circuit. Above about 75 or 80 percent, however, occlusion trouble rapidly develops—those arteries have become the major elements of resistance to blood delivery to the heart muscle. The unfortunate consequence is that coronary artery disease is asymptomatic until it's quite advanced.

While on the subject of coronary arteries, two more points ought to be mentioned. First, they have a peculiarly part-time operation. When the heart squeezes in systole, these vessels get scrunched and thus occluded. So they fill during diastole, unlike other arteries, and that means that they're filled by the lower diastolic rather than the higher systolic blood pressure. Thus they normally work at pressures lower than those in other systemic vessels. Second, all the trouble we have with coronary arteries might have been avoided if we had our entire heart downstream from our oxygenating organs as is the systemic heart of cephalopods. With fully oxygenated blood in the heart, an octopus doesn't bother with much in the way of coronary arteries. The systolic squeeze of the heart forces blood out into the cardiac muscle, which has a rich capillary bed. Blood is then collected in veins on the outer surface of the heart and returned to the main circulation.

Designing for Economy

In the poem by Oliver Wendell Holmes[6] entitled "The Deacon's Masterpiece" a carriage is constructed so perfectly that no part is disproportionately durable. It lasts 100 years and a day and then; well, then "It went to pieces all at once, All at once, and nothing first, Just as bubbles do when they burst." If the evolutionary process could have aspirations, that would be an appropriate one. The issue can be stated in a more general fashion. A system might be designed for the best operating economy, designed in such a way that the cost of construction and operation is minimized by having no part better than it need be. While precious little evidence is available to suggest that nature actually achieves quite such optimization, the notion at least marks a target at which natural selection takes aim.

Does the system of pipes in a circulatory system approximate such an ideal? For some features, the optimal characteristics can be calculated and compared with reality. Consider the branching of blood vessels and the cost of pumping blood through them. In every vessel the speed of flow is zero at the wall and maximum in the center—that's the consequence of the no-slip condition. Fluid is thus always being deformed, and viscosity resists that, the more so as the fluid's average speed increases. And the Hagen-Poiseuille relationship gives the cost of that resistance in terms of pressure drop needed to drive the flow. Imagine that a large artery tapered down to a small capillary—by the principle of continuity the speed of flow in the capillary would be enormous. Capillaries being small, the cost (remember that fourth power of pipe diameter) would be out of sight, even assuming the capillary didn't explode. But we're obviously not built that way, with one aorta and roughly 7 billion capillaries.

Now imagine a more reasonable system, one in which each artery branches repeatedly into arterioles and capillaries. In this particular system the total cross-sectional areas of artery, of arterioles, and of capillaries are equal. Again applying the principle of continuity, we can readily see that the average speed of flow in each of the three should also be equal. That fourth power relationship still makes mischief—consider a simple bifurcation. If a pipe divides into two equal branches without change of total cross section, then it takes all of eight times (two cubed) as much work or energy to get fluid through a given length of daughter pipes as to get it through the parental one. In short, the cost of pumping fluid through the smaller pipes is still disproportionate, if not so catastrophically so.

Clearly a rational circulatory system should be arranged to branch in such a way that the cost of flow in daughter pipes is of the same order as the cost in parental pipes. Just as clearly, that will require the total cross-sectional area of the small pipes to exceed that of the large pipes. Of course, that's what we already figured out for our aorta and capillaries when calculating the number of capillaries—area increased 400-fold. In fact, one can go further and calculate a kind of optimum from the Hagen-Poiseuille equation, the deacon's beacon, so to speak. To do that, or at least to explain how it's done, requires that we know a little more about laminar flow.

Steady flow down a long pipe reflects a balance between two forces, a push and a resistance. The former, the pressure force, pushes lengthwise and acts across each cross section of the pipe. The latter, the viscous force, comes from the walls of the pipe. It results from the no-slip business and thus from the distortion of any chunk of fluid flowing along anywhere between the walls of the pipe and its center. Using the requirement that push and resistance must balance, it's no very difficult matter to calculate the way in which the speed of flow varies across the pipe, from any wall to the axis and over to the opposite wall. Figure 5.5 gives the outcome—the shape is called a *parabola*, perhaps a distant memory from a class in algebra. If one considers the whole two-dimensional distribution of speeds across a pipe, the shape is a *paraboloid of revolution*, of which the parabola is just a lengthwise, middle slice. By the way, the Hagen-Poiseuille equation can be derived from either—that's what I alluded to earlier when I mentioned that the equation could now be obtained theoretically.

(One might ask which Hagen-Poiseuille equation is the real one—the theoretical or the empirical one. Forced to choose, I'd opt for the

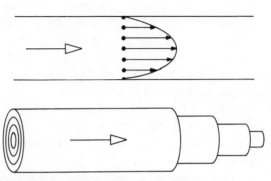

Figure 5.5. The parabolic distribution of the speeds of flow across a circular pipe, shown in cross section (above) and in three dimensions (below). In the latter, one gets a paraboloid of revolution, roughly approximated by a set of concentric cylinders.

empirical one, but that's just the bias of an experimentalist. In truth, both have value and limitations—getting a rule by making lots of measurements laces it to the real world, getting the rule by a theoretical derivation laces it to the larger corpus of science. That's why we're happiest when we can do both and, as here, get the same result. The distinction tends to get obscured in a world awash in computers, but it still exists. Indeed, one can recognize two fundamentally different uses of computers—for dealing with large bodies of data and for exploring the consequences of theories.)

Immediately adjacent to any wall, the speed of flow changes fairly steadily with distance from the wall. At the wall, the speed is zero; the further we look from the wall, the faster is the flow. We speak of that as a *speed gradient*—the way speed changes with distance, or, formally, a change in speed divided by the corresponding change in distance. Gradients, by the way, have a central and often unappreciated role in just about everything. The altitudinal gradient, or how distance up changes with distance across, determines how hard it is to climb a hill and whether water moving down hill will form a gentle brook or a raging torrent. A pressure gradient (pressure drop per unit length) has already been introduced. In the next chapters, concentration gradients will drive oxygen from capillaries to tissues. Temperature gradients determine, for example, how fast something (such as a roast) heats up or cools down, and lots more.

Our problem of figuring the cheapest arrangement of pipes turns out to involve nothing more nor less than calculating the relative dimensions of pipes so that the steepness of the speed gradient at all walls is the same. The calculation was done by Cecil D. Murray, of Bryn Mawr College, back in 1926, and is spoken of, when (uncommonly) it's mentioned, as "Murray's law."

Murray's law isn't especially complicated, and anyone with a hand calculator can easily play around with it (but you can ignore the specifics without missing the present message). The rule is that the cube of the radius of the parental vessel equals the sum of the cubes of the radii of the daughter vessels. If a pipe with a radius of two units splits into a pair of pipes, each of the pair ought to have a radius of about 1.6 units. (To check, cube 1.6 and then double the result—you get about 2 cubed.) The daughters are smaller, but only a little (Figure 5.6). Still, if the parental one eventually divides into a hundred progeny, the progeny do come out substantially smaller, with each about a fifth of the radius of the parent.[7] (Their aggregate cross-sectional area is, of course, greater than the parental one—to be specific, four-fold greater.)

The relationship predicts the relative sizes of both our arteries and our veins quite well. It only fails for the very smallest arterioles and capillaries. Still, that's not unreasonable. Murray's law was derived using a set of assumptions that are violated in those tiny vessels. When the bore of a pipe is about the same size as particles flowing through it, the Hagen-Poiseuille equation isn't applicable. The main problem is that the flow can't possibly have the parabolic distribution of speeds that the equation presumes. That's what happens in the tiniest vessels. Red blood cells just fit through, so flow is, on the average, faster near the walls and slower in the middle than would be the case for a normal, homogeneous fluid.

It would be indefensibly anthropocentric to suppose that we're the only creatures that follow Mr. Murray. My friend Michael LaBarbera (who introduced me to the whole issue) has tested the law on several systems that are very unlike us structurally and functionally, and very distant from us evolutionarily. These are sponges, which take water in through tiny pores (hence their biological name, *Porifera*), separate any microedibles from it, and pass the water through pipes of increasing size until finally disgorging it. Murray's law again proves applicable; it

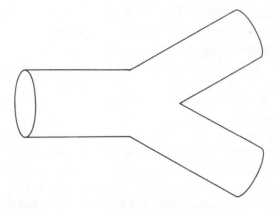

Figure 5.6. A bifurcation in a pipe drawn to follow Murray's law so the cost of flow in the daughter pipes is the same as that in the parental one.

even works for vessels measured on fossils of a long-extinct group of sponges. It again breaks down where the assumptions under which it was derived are violated. Thus in sponges it fails to predict the actual sizes of the vessels where filtration and propulsion are simultaneously accomplished. These vessels, however, must have very unusual patterns of flow since the cells that do the filtering and pumping also form the lining of the vessel walls.

If this were as far as we could push Murray's law, it would certainly be a fine generalization, another satisfying application of theory to unravelling biological design and another bit of evidence that physics matters. Recently, though, some tantalizing concomitants have emerged. It has been found that if flow is experimentally reduced in a blood vessel, the inside diameter of the vessel decreases until it stabilizes at a new and smaller size. The reduction in diameter is just about what one would expect if the system were rearranging itself to keep the speed gradient at the wall unchanged. And the mechanism for the corrective adjustment is becoming clear. Without getting into the details, it looks as if the cells lining blood vessels can quite literally sense changes in the speed gradient next to them. An increase in the speed of flow through a vessel increases the speed gradient at its walls. An increase in gradient stimulates cell division, which would increase vessel diameter as appropriate to offset the faster flow. Neither change in blood pressure nor cutting the nerve supply makes any difference—this is apparently a direct effect of the gradient on synthesis of some chemical signal by the cells. Perhaps the neatest feature of the scheme is that a cell needn't know anything about the size of the vessel of which it's a part. As a consequence of Murray's Law, it can be given the same specific instruction wherever it might be located, a command telling it to divide when the speed gradient exceeds a specific value.

Such changes in vessel size mean that we have an elegantly self-adjusting system. Murray's law works because the cells lining the vessels make it work. They get wider (or narrower) as required to maintain the average speed gradient at their walls at some standard value. They do this initially during fetal and juvenile growth, but at least the smaller ones apparently never give up that self-adjusting character. Chemical factors are known that promote the growth of blood vessels into areas of active metabolism; here's the complementary mechanism to adjust the sizes of those blood vessels. Compensation for changes due to growth, and so forth should be automatically elicited. Indeed, can you imagine recovering from even the most minor injury or surgery without such capability?

In fact the existence of these adjustments has long been presumed, even without recognition of a specific mechanism. I exercise to improve the blood supply to my heart. Greater activity makes the heart push the blood faster, which increases the speed gradient at the walls of the vessels, and which, in turn, stimulates the vessels to expand and increase their capacity for conveying blood. Blockage in one place stimulates expansion of vessels nearby, or in the jargon, causes improvement in the collateral circulation—a reasonable response if the blockage increases flow through those nearby vessels even a little.

Beyond that, we see a piece of a larger biological puzzle. A fertilized egg may have quite a bit of information in its DNA, but creatures such as ourselves are more than just quite a bit complicated. The question, then, is how we can possibly be made with, relatively speaking, a limited amount of genetic information. Clearly not all details of the assembly can be specifically read from the genetic blueprints. What certainly happens is a lot of local adjustment and self-organization, of which this system that puts blood vessels of proper sizes in proper places is as good an example as any. Putting the principle more pretentiously, we can recognize that properly contrived local responses can certainly produce large-scale optima. This kind of goal-directed self-assembly may strike us as strange because it's quite different from the scheme underlying most of our technology. Two cars off an assembly line perform similarly because they are similar in their structural details. Two organisms of the same species or even the same litter look and behave alike, but prove quite remarkably different in cellular and tissue-level detail.

It's so nice that we creatures pass the test for good design, at least with respect to the dimensions of the vessels of our circulatory systems! Nonetheless, I hasten to point out again that finding such fine design doesn't constitute evidence that evolution by natural selection is the generating mechanism. For that, one has to look elsewhere. As Stephen J. Gould points out, perfection in design is as likely to result from one system of design as another; it's the character of the imperfections and the improvisations that can shed light on origins, relationships, and mechanisms. On the other hand, finding *what we believe to be* good design isn't trivial either. It means that we biologists can have a little more confidence that we have some sense of what's important in the design of a particular functional device or system. *Apparent* poor design, as a rule, is most likely to point a finger at some serious inadequacy in our understanding of what's at issue. It's probably not nature that's poor, but rather our science.

Notes

1. Don't try to make your pet iguana dive. The diving ones here alluded to are the giant iguanas of the Galapagos Islands, off Peru.
2. Speed is distance over time; area is distance squared. So volume flow rate must be distance cubed over time. Whatever the units, whether gallons per minute, cubic meters per second, or any other, the underlying measure is still a volume (a distance cubed) divided by some time interval.
3. Reynolds was the first professor of engineering at Owens College, Manchester. The title of the great paper of present interest nicely gives its main thrust: "An experimental investigation of the circumstances which determine whether the motion of water shall be direct or sinuous."
4. By ruling of the Antiacrimonious Acronymical Accreditation Association.
5. The statement "is proportional to" is a little less specific than "is equal

to." It implies that if one thing is, say tripled, then the other is also; if one thing goes up tenfold (is multiplied by ten), then the other does as well. In short, the two variables follow, multiplicatively or divisively.

6. Holmes (1809–1894) was professor of anatomy and physiology at Harvard, so perhaps he's a particularly appropriate source for insight on the present subject.

7. Since $0.215 \times 0.215 \times 0.215 \times 100 = 1.00$.

6 *Why Move Blood?*

One way to appreciate the necessity of some functional arrangement is to consider the consequences of its absence. Without being explicit about it, we use that kind of reasoning all the time. Will the car give grief if the brakes aren't relined? Would life have been better if we hadn't adopted three kittens? What if I had bought Eclectic Electric at 5¼ or Flybynight Airlines at 75? Playing really elaborate "what if" games with one's personal circumstances might not improve your mental health, but playing them with human history must be at least a subliminal activity of every historian. At the very least, they provide practice in the use of the subjunctive mood.[1]

Anyway, this process of assuming some condition clearly at variance with reality and then examining what would change were it the case is quite a respectable procedure in the great game of science. We grandly call our questions *hypotheses* and recognize that as often as not they are posed as "what if it were otherwise." In a lecture about human population growth, I wanted to show that the current rate, of about 2 percent per annum, was a historical aberration. I wanted to show that even though our species was assuredly much less common a few hundred thousand years ago, we could not have grown at such a rate. So I assumed the 2 percent figure and looked at its implications, not for the uncertain future, but for a past of which we have some record. Specifically, I extrapolated back, with the aid of a simple computer, to find out how long humans must have been on the scene. The result, surprising even me, was that, metaphorically speaking, Adam and Eve left Eden only about 1100 years ago! It would take quite a revisionist historian to accommodate that datum, millennia shorter even then Bishop Ussher's celebrated biblical calculation. The conclusion is that a 2 percent historic norm doesn't wash.

"What if it were lacking," a look at the consequences of an omission, is merely a particular version of such a hypothetical headstand. At this point I want to view circulatory systems in just that way, by considering what life would be like without them. Now I don't mean what my life or yours would be like in any immediate sense—as they say in the shoot-'em-up murder mysteries, we'd be dead meat PDQ. What I mean is that lots of organisms lack any equipment at all for pushing any fluid from place to place within themselves, not just circulatory systems *sensu strictu*. I want to ask what forms they can't come in that are open to us, and what they can't do that we find possible.

Random Walks and Diffusion

The underlying issue, of course, is the business of moving material around inside organisms. It turns out that moving material by moving a fluid—moving stuff by what's called *bulk flow* or *convection*—is only one of two ways things are gotten around. Without a circulatory system, an organism is entirely dependent on the other mechanism, that of *diffusion*. So we have to say a little about the characteristics of diffusion as a physical process. In fact, it is one of the most peculiar and counterintuitive of physical process. It's not driven by any ordinary pump, or even, like the rise of hot air above an iron radiator, by a difference in temperature between two places. Instead, it's driven directly by heat, by the fact that with any heat around at all, that is, at any temperature above absolute zero ($-273°$ C), molecules of gas or liquid are in continuous, random movement. Each molecule moves until it hits another and then bounces off to move in another direction as a component of a kind of ideal cocktail party.

The usual analogy for diffusion is built on a somewhat less refined version of that cocktail party. One is asked to imagine a person at a given spot in a state of idealized (as opposed, I suppose, to ideal) inebriation and disorientation. The drunk takes a step, its direction quite random, gets disoriented again, takes another step in a direction quite unpredictable even if the previous direction is known, and so forth. Quite obviously the person moves besottedly around; quite as obviously one can't predict the direction from the starting point to wherever the character might be at any time. One does know a little more about location than about direction—as time goes on, it's at least possible to find the person further and further from the start. Still, a return to the starting point or even a march off in the opposite direction from the initial one is always at least possible. The drunk is, of course, the analog of a molecule of gas or liquid; the step is the motion of that molecule through space until it hits another; the process is called, simply, a *random walk*. Figure 6.1 gives an example of a random walk, in this case one generated by a computer.

Four main points will connect all this ethanolic ambulation with the phenomenon of diffusion. The first is that, despite one's inability to say much about where a single walker will be, increasingly accurate predictions of a statistical nature can be made as the number of walkers (= molecules) and the number of steps (= time) increases. Figure 6.2 gives the end points of a large number of walks, each involving the same number of steps (or, we might say, the same elapsed time). A certain statistical order indeed begins to emerge, with clustering around the starting gate and a diminishing population density further out. Atoms and molecules are exceedingly small, vastly smaller even than cells,[2] and they collide extremely often. So we're almost always interested in the group behavior of very large numbers of wandering entities over a very large number of steps. While it's hard to be specific about how a few molecules diffuse, there's little point in worrying about the matter. Conversely, it's important to anticipate what large numbers will do, and that's something about which quite definite (although ultimately statistical) rules can be derived.

The second point is that, at least in one sense, nothing goes anywhere in particular—diffusion is intrinsically omnidirectional. A cloud emerges in every direction from a point, as when some molecules from a tiny solid lump of sugar spread through a liquid medium. If directionality is apparent, it's evidence that something other than (or in addition to) diffusion is taking place. The usual something is convection.

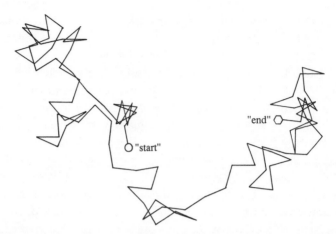

Figure 6.1. A random walk, starting at a given point, with steps of equal length. This one really is computer-generated, not an artist's impression. Almost no random walk ever really *looks* random!

Suspend a lump of sugar by a string in a container of *very* still water (actually not at all an easy thing to do), and the water at the bottom of the container will get sweet first. What happens is that the sugar-water around the lump is denser than the surrounding pure water, so it slowly sinks toward the bottom. This latter process is called *free convection,*[3] and it's most emphatically distinct from molecular diffusion. (As Trollope pointed out with astonishing prescience in *Ralph the Heir*, "Gentle readers, the physic is always beneath the sugar, hidden or unhidden." Or maybe he had something entirely different in mind.)

Diffusion certainly involves the movement of material, even though the movement is queerly omnidirectional, which brings us to the third point. The cloud of molecules does spread over time, and molecules of sugar can be detected at increasing distances from a lump, but this point is stranger and more subtle. Assume we have a situation involving some source, like our sugar lump, spreading purely by diffusion, and there's some minimum concentration of sugar in water that we can taste. The longer the time available for diffusion to take place, the further from the lump this delectable threshold will be. That much should be obvious. The strange part is that the time course of movement is like that of no motion with which we're intuitively familiar. In everyday

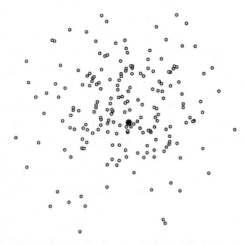

Figure 6.2. The end points of a large number of random walks generated by the same program (although not shown at the same scale).

motion, constant speed is our baseline; the assumption of constant speed means that an object can go twice as far if given twice the time. To go 100 times as far takes 100 times as long. In short, the time it takes is proportional to the distance covered. For the spread of a diffusing cloud of molecules the rule is different. The time required to reach some particular concentration is proportional, not directly to the distance from the source, but to the *square* of the distance from the source. To reach that concentration at twice a certain distance takes four times as long. To get it ten times as far away takes 100 times as long.

This diffusional domain is indeed a weird world. It is as if a trucking company, finding that it has to carry stuff twice as far, needs to have four times as many trucks working. In effect, speed decreases as distance increases, at least if we use our ordinary notion of speed. In fact, any talk about the speed or rate of diffusion tacitly uses quite an unusual version of the notion of "rate," something one has to bear in mind.

The fourth point is that we're talking only about *net* movement when worrying about reaching thresholds, trucks, and so forth. Remember that the drunk was free to move back toward the starting point and quite often did so. What, then, determines the apparent direction of purely diffusional motion? Again, it's just a statistical matter. At any time more sugar molecules are near the lump than more distant from it. When they are near the lump, more directions lead away from the lump than toward it. So, on the average, more molecules will move away from it than toward it. Thus diffusional motion, on the average, causes net movement from areas of higher concentration of whatever's diffusing to areas of lower concentration. All else being equal, the net amount diffusing is typically proportional to the difference in concentration between two locations. Or, if we consider distance as well, it's proportional to the concentration difference divided by the distance between the locations—the concentration gradient, as mentioned in the last chapter.

For present purposes, the most important point about diffusion is that third one about rate. Diffusion is an excellent way to move molecules for short distances. For example, when an impulse reaches the end of a nerve and has to initiate another impulse in the next nerve, a diffusional link most often intervenes. A substance is liberated by the first nerve and diffuses across to the second. The distance is about 20 nanometers, a little less than a millionth of an inch, and the process takes only about a thousandth of a second. Conversely, diffusion is a terrible way to move molecules over long distances. The times needed go up with the square of distance, not directly with distance. Knut Schmidt-Nielsen gives a nice illustration. Say it takes a hundredth of a second for oxygen to diffuse the typical distance of a hundredth of a millimeter between a cell and a capillary. If the distance were 1 millimeter instead, then it would take not 1 second, but 100 seconds, 100×100 times longer. If the distance were 1 meter (roughly the distance between lung and hand), then it would take 100 million seconds, or about 3 years.

Matters are (or one might say matter moves) somewhat better in gases than in liquids—a moving molecule gets further between collisions in a less dense medium. Thus the measure of the speed of diffusion for oxygen is about 10,000 times higher if the oxygen is in air than if it is in water or blood. The relative sizes of the alveoli of the lungs (the final air chambers) and the capillaries in the alveolar walls almost certainly reflect just that difference. One can predict that to balance the design the alveoli should be larger than the capillaries by a factor that is the square root of 10,000, or 100. Alveolar size is a bit indeterminate, since alveoli inflate and deflate with every breath, but the predicted value is not far from reality.

Even in gases, however, diffusion is a slow business if distances are appreciable. That's part of the reason why contriving a decent demonstration of diffusion is surprisingly tricky. The other part is the necessity of holding convection at arm's length. Drop a crystal of dye or a drop of liquid colorant into water, and the color spreads around, but convective bulk flow is at least as responsible as is diffusion. Open a bottle of perfume in a room, and pretty soon everyone gets a whiff— here bulk flow is overwhelmingly predominant, with diffusion important only in the last fraction of a millimeter next to the walls of your nasal passages. To get a decent demonstration of diffusion you have to stabilize some water so it can't convect around in the container. The easiest way to do that is to add about 1.5 percent gelatin, as directed on the package from the grocery. After solidification in a glass jar pour some colored water over the gelatin, return it to the refrigerator, and observe, not after a few minutes, but over several days.

Having wandered back into the kitchen, a further digression is irresistible. Heat transfer in solids follows the same rules as does diffusion in liquids and gases, so the former provides a convenient conceptual analog for the latter. As distance increases, it takes a disproportionately long time for heat to penetrate. If you cook two roasts, one twice as thick as the other, the thicker one ought to take, not twice as long to cook, but four times as long—if the criterion of doneness is achievement of a certain temperature in the center. Amusingly, the cookbooks commonly assume an even more disproportionate relationship—not time proportional to distance squared, but to distance cubed. At least that's the implication of giving data for cooking times relative to weight. (Weights are proportional to volumes, and volumes are proportional to the cubes of distances.) So in effect the cookbooks suggest that the larger roast be cooked not twice, not four times, but all of eight times (two cubed) as long as the smaller. Still, the best authorities recognize that a problem exists. The average cook couldn't cope with instructions to adjust times in proportion to the two-thirds power of weight, so they just give a few precautionary words. James Beard notes, "The smaller the roast, the longer the time per pound," while Fanny Farmer advises "not more than three hours even for a very large roast."

A different rule, by the way, should apply for heating a container of liquid. Put a set of pots of water of different sizes in an oven and determine how long each takes to reach a given internal temperature. The pots will heat water in proportion not to diameter squared, like the roast, but to diameter directly. What happens is that the water convects itself around as a result of the initial jostling and of variations in temperature. Once heat has penetrated the surface of the pot, it is for practical purposes immediately mixed around. That, by the way, is analogous to what goes on in many plant cells, in which the contents circulate around in a process (you may recall from some name-ridden biology course) called *cyclosis*. Plant cells are larger than animals cells, not, I think, incidentally, as you'll see from the next stage of our argument.

Why, Then, Circulate?

Our initial question, the one that got us involved in this rather lengthy discussion of diffusion, asked about what life would be like without circulatory systems. An answer is now apparent. It would be very hard to imagine any large, active, and well-integrated organism without some system of bulk fluid movement to augment diffusion. For the kinds of creatures we're considering in this book, macroscopic animals, circulatory systems are the principal bulk fluid transport systems. Assume no circulation, and we are forced to discard ourselves and all like us as quite impractically large.

That isn't to say that in banning circulatory systems we've entered the realm of science fiction. Far from it. As mentioned, an enormous and diverse host of organisms manage to stay in circulation without such systems. As far as we know, all meet a particular criterion—they present only short distances over which dissolved molecules have to be moved. The criterion can be met in several ways, though, which is probably why that host is so diverse. First, the creatures can be small—of the sizes of cells, and for the same reason that animal cells are no more than that size (about which more a few pages ahead), very roughly a hundredth of a millimeter in length. Alternatively, the creatures can be flat or threadlike, as are many algae. Or the metabolically active parts of the creatures can be limited to an outer layer on a fairly inert core. The latter is what many very large algae and the bigger sea-anemones and jellyfish do. Still, even if clearly not disabling, these are severe-limitations.

This whole discussion allows us to state what might be taken as a functional definition of a circulatory system, although it includes a number of arrangements not normally subsumed under that rubric. These systems, then are, *devices that use some form of convective bulk flow to reduce the effective distance between two parts of an organism or between the organism and its environment.* They are, to reiterate, schemes to get larger without constraint by that terrible distance–time relationship of diffusion.

I hasten to emphasize that circulatory systems do not truly replace diffusion with convection; what they do is to augment one with the other. In every arrangement to transport material from place to place with a circulatory system a diffusional component always remains. Material gets from capillaries to cells mainly by diffusion. We forcibly inhale air and odorants, but the odorants cross into our sensory cells by diffusion. Oxygen diffuses from alveoli to capillaries in the lungs. In short, diffusion is always retained for short-distance transport.

The Design of a Circulation

The chapter began with talk about physical imperatives and constraints. Retaining that outlook, what more can we say in general terms about how circulatory systems ought to be arranged? In the last chapter we noted that the cross-sectional area of our systemic capillaries exceeded that of the aorta by a factor of 400. With small capillaries and a large aorta, that was the only way flow could be slow enough in the capillaries to keep the cost of pumping blood through them from being extravagantly high. Slow flow has an additional virtue in addition to keeping down cost—the greater the time it takes blood to pass along the length of a capillary, the greater the time available for diffusive

exchange between blood and surrounding tissue. The general arrangement of large pipes with fast flow and small pipes with slow flow has
yet another benefit. Again by the principle of continuity, fast flow has
the effect of keeping the volume of blood contained in those large pipes
to a minimum. That's a little subtle, but you'll see what's going on
when you realize that fast flow means that the big pipes need only be
big, not huge. What's happening is that they can deliver a lot of blood
relative to their cross-sectional areas in a given time, and thus the total
volume of blood in the large vessels can be relatively small.

The basic arrangement of our circulatory system, then, turns out to
be completely general, characterizing every system we know of in nature in which a gas or liquid is moved around by some bulk transport,
convective scheme. Both wide pipes and narrow pipes are used, and the
cross-sectional areas of the narrow pipes exceed that of the large pipes.
The narrow pipes are very short and used only at sites of material exchange such as capillaries. The wide pipes are relatively long and are
used for transport between one exchange site and another. For the few
systems where we have sufficiently detailed information, Murray's law
applies to the relative sizes of the pipes.

A proper skeptic might argue that any distributional manifold of
pipes would at least approximately follow these general rules simply as
a geometric consequence of going out to all places from one starting
point. The skeptical argument, in short, is that we're reading too much
into nature's design. Maybe a road system leading outward from a city,
in which the number of lanes reflected the traffic level, would take
much the same form. Luckily, we're in a position to apply the now
familiar "what if it were otherwise" logic to the matter. Back in the
1920s, August Krogh (already mentioned as the first person explicitly
curious about circulation in giraffes) made careful measurements on a
distributional system of pipes in which transport was entirely diffusive,
one with no bulk flow at all. Recall from Chapter 1 that insects send
oxygen out to their cells through a set of air-filled pipes, using a tracheal system very roughly analogous to our respiratory air pipes. As
mentioned, diffusion in air is far faster than diffusion in any liquid, so
the arrangement isn't intrinsically primitive or impractical. Some insects, especially in flight, use pumping devices to propel that air, but
Krogh worked on a caterpillar that didn't pump. What he found was

that total cross-sectional areas in the tracheal system did not increase as pipes divided into ever-finer branches. Rather, total cross-sectional area was just about the same at any level in the system—the optimum arrangement for a wholly diffusional transport system. So an increasing cross section as narrower pipes are reached isn't just some geometric imperative. Where nature uses a different bit of physics, she adopts a different design.

The Size of the Whole Versus the Size of the Parts

As organisms go, we're big ones. Organisms range in length from a few micrometers (thousandths of a millimeter) to a few tens of meters. It's revealing to look at that range on a geometric scale, that is, looking not arithmetically at *how much* bigger one creature is than another, but geometrically at *how many times* bigger it is. We commonly call each tenfold increase an *order of magnitude*, so the length range of organisms, 10 million-fold, is seven orders of magnitude. Now humans are about 2 meters in length, so we're only one order of magnitude from the largest but all of six orders from the smallest. On such a geometric scale a typical organism would be about 1 centimeter long, a little less than half an inch. This brings up two points. First, this is a wide size range; and second, we're substantially atypical, too corpulent by far.

The larger point (beyond our own largeness, to which we'll return), implicit but important in all the preceding talk, is that what constitutes a good design is very much a matter of the size of an organism. When one talks about such matters as circulation, size takes on an importance that transcends minor matters of ancestry, feeding habits, or even whether a creature lives in air or water. With circulation, again, what's absolutely crucial for a large creature would be redundant baggage for a very small one. In general, what constitutes practical reality depends very much on how big you are. Ants lift disproportionately large stones with the same sort of muscles we have[4]; but, however combative, ants never throw stones. An ant-sized creature just can't impart enough momentum to a projectile to do damage to another. I clean out my rain gutters with care, deliberation, and sobriety lest I fall, but no baby bird is likely to suffer injury by falling from a nest. And, to take a wild extreme, Edward Purcell, a physicist, did a calculation showing that a bacterium needn't bother to swim around if it lives in a uniform broth of nutrients. At its size, diffusion will provide provisions at a rate sufficient even for its voracious metabolism.

Cells, whether free-living or parts of animals, are very roughly a hundredth of a millimeter in length, a size at which diffusion is a potent agency indeed. Plant cells are typically an order of magnitude larger, a tenth of a millimeter. As mentioned already, that larger size is almost certainly related to the presence in plant cells of an arrangement to keep things stirred up. Beside that, many (I don't know whether one can say most) plant cells have their metabolically active part just adjacent to the outer wall, with the middle occupied by a fairly inert bag of liquid. Putting aside both plant cells (on one hand) and such things as large sea anemones (on the other), it appears as if the cellular domain is diffusion based, while the multicellular domain augments diffusion with convective bulk flow, or circulation. That cell size is nearly constant despite the great variability of organism size probably reflects both the peculiar time–distance characteristic of diffusion and its general slowness.

A decently large body of information exists concerning the sizes, speeds, and capacities of the various components of circulatory systems, although the distribution of data among animals has the strong odor of anthropocentrism. That is, more data exist for mammals than for anything else, followed by birds, cold-blooded vertebrates, and then all other animals. I'll go through a little of the data, partly to illustrate how a comparative viewpoint gives a person a little better perspective on how a system functions and partly to show how a lot of numbers can add up to a non-numerical picture. Much of the information I'm using comes from an elegant book (*Scaling, or Why Animal Size Is Important*) by my colleague, Knut Schmidt-Nielsen, the one who produced those nice comparisons of diffusion over different distances. Incidentally, half a century ago he was a student of August Krogh, the father of comparative physiology, some of whose concerns you've already encountered.

First, a bit of what is now classical knowledge. It's possible (combining the ingenuity of Daedalus with the patience of Job) to measure the rate at which inactive animals consume oxygen or produce heat, obtaining data called *resting metabolic rates*. That these rates are higher in large mammals than in small ones comes as no surprise. Less self-evident is the fact that the metabolic rate of large mammals *relative to their body weights* is *lower* than that of small ones. A gram of elephant is less active, metabolically, than a gram of mouse. The difference of about 20-fold certainly isn't trivial. Even their individual cells, although the same size, look different, with those of the mouse having far more machinery packed in. Not that the lower relative rates of large mammals are truly paradoxical—consider the difference in deportment between large and small ones: The large ones are phlegmatic, and the small ones frenetic, with the latter, for instance, eating far more relative to body weight.

Conclusions drawn from these resting rates have been quite reasonably criticized as not reflecting the real demands placed on the systems, but recently data have been obtained on maximum metabolic rates from animals working steadily, mainly on treadmills. It seems that for mammals as a whole, the resting rates can be increased about tenfold in sustained activity, and that figure of tenfold doesn't vary much between small ones and large ones. So the old picture that small mammals live much more intense lives is still very much with us. Elephants live longer than mice; in general large animals live longer than small ones. The possible connection of this latter phenomenon with the differences in relative metabolic rate has been the subject of continued debate and speculation. Unfortunately no consensus has emerged on what might link lifespan and relative metabolic rate.

Metabolic rate is just one of the variables that change in a systematic way over the size range of mammals. Large mammals have relatively smaller brains, eyeballs, kidneys, and livers. On the other hand, they have heavier bones unless, as with whales, they're aquatic. Thus a mouse-sized mammal is about 4.5 percent skeleton, a cat-sized mammal is about 7 percent; while an elephant has a little over 12 percent of its weight invested in bone. Despite the extra supportive equipment, the large mammals are relatively more fragile, but that's a tale for another book. The point to be made here is that systematic variation is typical, and constancy in relative size over a range of body sizes is less common and therefore noteworthy.

Thus the existence of a series of interrelated items that don't vary with animal size might have something to tell us. Turning back again to features of circulatory systems, we can note the following.

1. Red blood cells are about the same size in all mammals (although they're larger in some other vertebrates, most notably amphibians).
2. The viscosity of blood is about the same (see 3).
3. The hemoglobin concentration (about 130 grams per liter) and the fraction taken up by red blood cells (about 45 percent) don't vary a lot. That constancy, as noted in the last chapter, probably reflects the steep rise in viscosity if more cells are packed in and a consequent uneconomic extra load on the heart or slowing of flow.
4. The amount of blood in the body relative to body weight is nearly constant at 60–70 cubic centimeters per kilogram or 6–7 percent of body weight.
5. The relative weight of the heart also varies little, remaining at about 0.6 percent of body weight. We all have about eight or nine times as much blood as heart, about which datum one is free to wax metaphorical, metaphysical, or sociological.
6. Blood pressure doesn't vary in any systematic way with size, either, except for being higher in especially tall mammals.

Obviously we shouldn't expect every relevant circulatory quantity to be size-independent. Somewhere we ought to find variation concomitant with the size-dependence of relative metabolic rate. After all, more oxygen must be made available to the cells of small mammals. Oxygen must diffuse from capillaries to cells for the cells to do their stuff, and a mouse cell clearly does a lot more stuff per unit time than does an elephant cell. Does this greater need call forth some functional difference in circulation? There seem to be two important differences at this level of the system. First, the capillaries of small mammals are closer together, so no cell is as far from a capillary in a mouse as in an elephant. Second, the hemoglobin of the blood of small animals, while it holds as much oxygen, holds it less tightly. Thus a higher concentration of oxygen is available for diffusion out of the capillaries. (Mouse hemoglobin and elephant hemoglobin are slightly different molecules.) Both of these differences independently increase the steepness of oxygen's concentration gradient between capillaries and cells and thus the efficacy of diffusive supply.

A similar situation emerges when we look at hearts and blood volumes. Why so much of either in big mammals if they have relatively lower metabolic rates? In fact, a satisfyingly logical answer is available. A bigger heart pumps more blood, but it doesn't pump blood very much faster. That's because the bigger heart beats less often. An elephant's heart beats around thirty times a minute, while a mouse's pulse rate may exceed 500. Since the speed of flow is the same, it takes blood longer to get where it's going in an elephant. On the average, a bit of blood takes about 2.5 minutes to make the full circuit of heart to body to heart to lungs to heart in an elephant. The same circuit takes only about 7 seconds in a mouse. (Human resting circulation time is about 50 or 60 seconds, as mentioned back in Chapter 1.) This means that the mouse is using its blood more often, loading each bit with oxygen in the lungs and unloading at the systemic capillaries about twenty times as often. Again that figure of twenty—the mouse's relative metabolic rate is twenty times that of the elephant, their rates of heartbeat differ by about the same factor, and now the utilization rate of blood differs similarly. Very tidy!

Mice, of course, aren't the smallest mammals, nor are elephants the largest. But the more extreme creatures are a little aberrant and therefore not quite as handy for comparisons. Shrews may be as much as ten times less massive than our paradigmatic mouse, but they are exceptionally active (and hard to tame) and have slightly unusual hearts in ways and for reasons not yet fully clear. Whales have the complication of being aquatic along with an obvious awkwardness as experimental subjects. It ought to be mentioned, though, that the story for birds is similar to that for mammals, with slightly bigger hearts but slightly slower heartbeat rates, and thus with comparable cardiac outputs. Hummingbirds are about as small as shrews and are similarly a bit aberrant.

Looking at these matters in a more general context, we might reasonably ask just what's ancestral or ordinary and what's derived or specialized when considering organisms of a range of sizes. Our own large size gives us an inadvertent bias toward regarding largeness as normal. The world of a relatively large organism is the world with which we're more likely to be comfortably familiar. Besides, it's the big stuff in museums that draw our attention. Brontosaurus always wins over any foraminiferan.[5] The bias, quite evident in much of the older paleontological literature, is an unfortunate one. It's increasingly evident that small size is more often ancestral and large size derivative. Early mammals were fairly small, as were early dinosaurs; large creatures are more often the evolutionary dead-ends—major evolutionary innovations seem to have happened far more commonly among smallish creatures.

(The most notable exceptions are among flying animals—insects, pterosaurs, birds, bats. In these the tiny ones are the most recently evolved. The best explanation brings us back to one of the main drums I beat, namely, the notion that there's an inescapable physical substratum with which life must contend. Active flight is quite a fancy activity, and lineages that have fliers seem to have begun with purely or predominantly gliding forms. For aerodynamic reasons related to the business of Reynolds number, mentioned in the last chapter, very small gliders don't work well. If gliding is the ancestral condition, then it's reasonable for ancestral forms to be fairly large. Conversely, only flappers work decently if small, and the completely obligate flappers, such as hummingbirds, microchiropteran bats, and flies and bees, are the relative newcomers. Giant fossil fliers aren't monstrous exaggerations of the media, just largely so.)

The size-related bias of people is an aspect of a large problem that bears at least occasional examination. Our science is subtly but pervasively corrupted by anthropocentricism. To the extent that the scientific description of the world reflects such hominid bias, it must constitute a less powerful and general description. Consciously purging that bias will always be necessary, since every scientist—past, present, and future—is unarguably human. The problem, it seems to me, is a far knottier and more intractable one than that of specific cultural biases, to which far more attention is currently being given. To make matters worse, this necessary campaign against anthropocentricism has an adventitious side-effect—it's yet another factor making science look cold, inhuman, and forbidding to nonscientists. As people we're only human, and the attempt to circumvent that awkwardness, however defensible intellectually, makes science less friendly, less immediately appealing, and less comfortable.

Notes

1. As in "Dear, I wish I were in a subjunctive mood."

2. After discovering quite accidentally that students tended to lump molecules and cells into the same mental box labeled "invisibly small," I tried to put some meaning on the difference in the two domains. The best I've come up with is a calculation that says there are about as many molecules in a cell as there are cells in, say, a cat.

3. As opposed to *forced convection*, which while also bulk fluid movement, is driven by some pump or fan rather than by differences in density within the fluid.

4. But they don't lift them far—so while they do better on force, they do worse on distance, and the work done (defined as force times distance) isn't exceptional.

5. Foraminifera, of which you've probably never heard, are shelled protozoa that have left copious fossils; they are the very stuff of the white cliffs of Dover. As useful markers for petroleum deposits, their occurrence is quite a practical matter, by contrast with creatures of merely academic and cultural interest such as dinosaurs.

7 *Pliant Pipes*

In most communication systems the least interesting components are the actual conduits. If you wire up a house, the main decision about the wires concerns their gauges, which you pick to handle the maximum current anticipated. The circulatory equivalent of that decision was touched upon when we talked about pipe diameters and Murray's law. In most electronic circuits, the physical layout of the wires is a trivial matter, chosen mainly for convenience of assembly. Only what's attached to the ends of those wires and the operation of those active components matters. In a plumbing system unnecessarily long conduits and sharp corners ought to be avoided since fluids have viscosity even if electrical currents don't. But besides gauge and route, what's left to worry the designer? In the design of the pipes of a circulatory system, as we'll see, quite a lot. As we'll also see, the details of their mechanical properties are exquisitely attuned to the various tasks they perform. To heighten interest in the story, these other features reflect some really subtle problems associated with nature's general use of nonrigid materials.

Why Arteries Must Stretch

The initial bit of trickiness is a result of the pulsatile character of a pumping heart. Pressure in the left ventricle varies from near zero to the maximum found anywhere in the system (thereby ignoring, for the sake of a strong statement, any hydrostatic increase below the heart). Flow rate follows pressure, both intuitively and according to the Hagen-Poiseuille equation. Thus, if our pipes were rigid, pressures elsewhere in the system would faithfully mirror those of the left ventricle, and flow would be as unsmooth as traffic on a street with stop signs at every corner. Actually, with a set of fully rigid vessels, the system couldn't function at all. In its systolic squeeze, the chambers of the heart get smaller; and, as we've seen, liquids don't easily compress. So either the heart couldn't eject any blood or else something would have to blow out somewhere in the circulatory system. Something else in the system simply has to get more capacious if the heart gets less so.

Of course, one might imagine a scheme in which pulmonary and systemic sides of the heart beat alternately; but in both mammals and birds the ventricles are formed of a single, functional muscle. The whole heart, for that matter, is triggered to beat from a single initiating event. Alternative beating is at least an anatomical possibility in squids and octopuses, but I haven't heard that even they bother with it. What really happens in both vertebrates and cephalopods, and indeed in all animals that push blood around through pipes, is that the pipes are flexible. Not only do they bend around corners and shift location as we move our appendages, but they're most emphatically not a system with a constant internal volume. Arteries in particular swell and shrink with every beat of the heart, in opposite phase to that swelling and shrinking of the heart of which Harvey made much. These, again, are the pulsations you feel whenever you lightly touch the skin just over the course of an artery.

Having excessively rigid pipes isn't at all a good thing, and it afflicts no small number of us humans. The signature of the condition is precisely what you're now in a position to guess. The difference between systolic and diastolic pressure in the large arteries is abnormally great. The system lacks what we call sufficient "compliance." If it takes a certain average pressure to push blood through capillaries at an adequate rate, it then takes a higher systolic pressure to achieve that average pressure. So having one's resting blood pressure measured gives a quick indication of whether one is afflicted with that unhealthy condition called *atherosclerosis* (from the Greek "scleros," meaning "hard").

Having arterial pipes whose walls can stretch and shrink at the relatively modest pressures we produce is thus a necessary device to smooth out pressure and flow in the system. The idea is nothing new—the way arteries acted to store blood during systole and then eject it during diastole was described in 1733 by the English clergyman, Stephen Hales. (Hales, incidentally, was the first person to measure blood pressure, among his very numerous discoveries and inventions. Anglican sinecures for people like him may have been an inadvertent system of government grants for intellectual activities.)

Anyone who does a little electronic tinkering will recognize that this smoothing is the action of a component called a *capacitor,* something you might connect across a switch so the current doesn't go on and off quite so suddenly. Otto Frank, in 1899, modeled this arterial capacitance with what he called a "Windkessel" or air chamber, an element somewhat like a balloon whose volume is proportional to the pressure within it. That's not unlike an air chamber incorporated into a household plumbing system to keep the pipes from banging when the water is suddenly turned on or off. Figure 7.1 shows several of these capacitive schemes. The point is that extensible arteries are a must. I reiterate this because we'll turn shortly to what looks like a different topic, but that will eventually bring us back to that crucial lack of rigidity.

Merely saying that something is extensible isn't sufficient specification. Stretch a rubber band, and it wants to snap back to its original length; stretch some chewing gum, and it quite happily adjusts to the new length, forgetting that it was ever short. The chewing gum suffers what's called *creep* or *stress-relaxation,* and it would be a poor material for an arterial wall that has to return to its original diameter around 3 billion times. A little creep can be a good thing on occasion, but this is simply not one of those occasions.

Nor is that all. Some things can extend and recover their original lengths, but they don't recover as forcefully as they were stretched. Some energy has been lost in the process, and we say they are low in a property called *resilience.* A dropped tennis ball returns to sphericity after a bounce, but a new and more resilient one returns more nearly to its original height as well. That indicates that less energy has leaked away in the bounce. The silk of the orb web of a spider is both strong and extensible, but it has very low resilience. The unusual com-

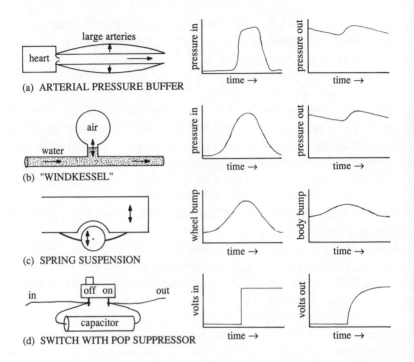

Figure 7.1. A stretchy artery, a "windkessel," and two other schemes for smoothing pulsations or making changes less abrupt.

bination of properties is most appropriate, since the net must be cheap and minimally visible (hence strong), it must decelerate flying prey gradually (hence extensible), but the prey must not be flung out again, trampoline-style, so low resilience is a virtue. By contrast, high resilience is clearly advantageous for an arterial wall since anything less than perfect resilience has to be paid for in energy delivered by the heart.

The Law of Laplace

Consider what happens when you blow up a balloon.[1] You apply pressure inside, which is to say that you create a pressure difference between inside and out. The balloon responds by stretching its walls and getting bigger. You apply *pressure to the central cavity* and as a result the *tension in the extensible membrane* increases. Pressure and tension are really quite different quantities. At the very least, pressure always pushes outward, while tension stretches a material lengthwise. Still, it would certainly seem reasonable to suppose that tension follows in direct proportion to pressure. That turns out only to work in the special case where the size of the object being pressurized doesn't change much. Apply pressure to an automobile tire, and the tension in the walls increases very nearly in proportion. Tires, however, aren't especially extensible—we reinforce their rubber with strands of steel or polyester cord of essentially constant length. Balloons, of course, do stretch; that's the whole point of them. Stretching complicates matters. The connection between applied pressure and the resulting tension depends on the size of the system, and in balloons that size varies as they're inflated.

We thus have three variables—pressure, tension, and size. Laplace's law, named for the nineteenth-century French mathematician and astronomer Pierre Simon Laplace but derived almost a century earlier by one of the Bernoullis, gives the relationship among the three for spheres or cylinders. By Laplace's law, tension is proportional, not to pressure per se, but to pressure multiplied by the diameter of the sphere or cylinder. Back, now, to what happens when you blow up a balloon. The tension you produce in the balloon's wall depends on the size of the balloon. For a small balloon a lot of pressure is needed to produce a given amount of tension, while for a large balloon it takes less pressure to generate the tension. Try blowing up spherical balloons of different advertised final sizes—you'll find the smaller ones harder to inflate! It's noticeably more difficult to blow up a 5-inch balloon than a 9- or 12-inch one.

The rule also applies to the problem of inflating a single balloon. To stretch a rubber band or an uninflated cylindrical balloon, increasing force is needed as it gets longer. Inflate a balloon and one gets quite a different impression. About the same pressure is needed throughout the inflation, except for an extra bit to get started and (if you persist) another extra bit just before the final explosion. The difference is that inflation stretches the balloon only indirectly, by applying pressure. As noted, pressure gets more effective in generating tension—stretch—as the balloon gets bigger, automatically providing the extra force needed as the rubber is extended. Partially inflate a cylindrical balloon and examine the stiffness of inflated and uninflated regions. The pressure inside is the same everywhere since no partitions intervene, but the expanded region is tense, while the unexpanded region is flaccid—same pressure, different tensions because of the different diameters.

An amusing sidelight on the phenomenon is given in the otherwise staid medical physiology text of Ruch and Patton. If one gradually inflates the urinary bladder of some experimental mammal, then the pressure in the bladder doesn't increase; rather, it remains nearly constant throughout the procedure. Apparently a substantial number of articles exists in the medical literature "explaining" this supposedly anomalous constancy, invoking such agencies as "reflex softening" of the bladder walls. A little undignified messing with balloons would have gone a long way toward clarifying matters!

Laplace's law explains all sorts of otherwise paradoxical observations. An automobile tire is decently inflated at under two atmospheres, at a pressure of around 26 pounds per square inch. The skinny tire of a racing bicycle, thin and puny by comparison, withstands (and, indeed, requires) four times that pressure. Why? Mainly because, due to the lower diameter of the tube, the walls of the bicycle tire are subjected to much less tension in spite of the higher pressure. Tiny plant cells, with walls no more than a thousandth of a millimeter thick, withstand pressure differences of many atmospheres. The walls are strong, but not exceptionally so; rather, the tensions to which they're subjected remain fairly ordinary despite the high pressures—since the cells are of such small diameters. If you buy pipe for household plumbing rated to withstand a certain pressure, you'll find that the wider gauges have thicker walls and are therefore much more expensive.

The relationship comes from purely geometrical considerations, quite independent of the material of the sphere or cylinder—we merely inquired as to how one material, rubber, responded. Whether the material stretches a lot before bursting or whether it splits its seams without much change in diameter is irrelevant to the applicability of the Laplace relationship. Thus we predict that big pipes need thicker walls than small ones whether these are rubber garden hoses, plastic sewer pipes, or copper water pipes.

Or, for that matter, blood vessels. The wall of the aorta of a dog is

about 0.6 millimeters thick, while the wall of the artery leading to the
head is only half of that. Pressure differences across the wall are the
same (recall Chapter 4), but the aorta has an internal diameter three
times as great. The bigger vessel simply needs a thicker wall. An arter-
iole is 100 times skinnier yet; its wall is fifteen times thinner than that
of the artery going to the head. A capillary is eight times still skinnier,
with walls another twenty times thinner. The practical proportionality
between pipe diameter and wall thickness isn't a very precise one, but
that's not surprising in view of all the other variables in these systems.
For one thing, the structure and mechanical properties of the walls vary
from big to small vessels; for another, by the time blood reaches the
capillaries, the pressure has dropped significantly. Nonetheless, a general rule that wall thickness is proportional to vessel diameter is clearly
evident, just the relationship expected from Laplace's law.

Veins sustain much lower pressures than do arteries, and their mechanical properties are, not unexpectantly, different. Still, big veins have
much thicker walls than do small ones, just like arteries. According to
the same body of doggy data just used, the inferior vena cava has a
diameter of 1 centimeter and a wall about 0.15 millimeters thick. A
venule is 250 times narrower inside; its wall is seventy-five times thinner.

How Arteries Must Stretch

I emphasized that our circulation was critically dependent upon having
arteries that are extensible, noncreepy, and resilient. They must smooth
the pressure fluctuations of a pulsating heart, complement the volume
changes of the heartbeat, and do both without using up unnecessary
energy. But such a combination of properties is still not good enough,
as we can see by turning again to balloons. Pressure, we noted, is more
effective in generating tension in the walls of a big cylinder than in
those of a small cylinder. Blow into a cylindrical balloon, and one part
of the balloon will inflate almost fully before the remainder expands.
Pressure inside at any instant is the same everywhere, but the responsive stretching is curiously irregular. What is happening is that any part
partially inflated is easier to inflate further than any part not yet inflated

at all. In fact, if rubber sheet didn't develop a little extra stiffness just before breaking, cylindrical rubber balloons wouldn't even be possible!

Premature expansion of one region is not at all what well-behaved arteries should do. What playing with balloons shows is that such premature expansion is what one would normally expect. So the arteries are doing something special, something for which cylindrical balloons are not adequate models. Their behavior is quite the opposite of that of urinary bladders, which were thought to be doing something special when they were actually behaving like a most ordinary elastic material. If the arteries weren't somehow special, you'd have a local expansion, an aneurysm, somewhere during every systole. Aneurysms, usually of arteries, do happen, and they're seriously pathological. About the worst place to have one is in your aorta, and the next worst is in the cerebral artery at the base of the brain. The problem, of course, isn't the aneurysm itself; it's the very high mortality among cases in which the aneurysm ruptures. Aortic aneurysms, at least, are commonly noisy, drawing attention to the problem. A graphic description of a terminally afflicted person is given near the end of the first Sherlock Holmes story, *A Study in Scarlet,* by Sir Arthur Conan Doyle.[2] Doyle, as it happened, was a physician, and so he probably wrote from clinical experience. (The "I", of course, is Dr. Watson.)

> "Then put your hand here," he said, with a smile, motioning with his manacled wrists towards his chest.
> I did so; and became at once conscious of an extraordinary throbbing and commotion which was going on inside. The walls of his chest seemed to thrill and quiver as a frail building would do inside when some powerful engine was at work. In the silence of the room I could hear a dull humming and buzzing noise which proceeded from the same source.
> "Why," I cried, "you have an aortic aneurism!"

The primary question isn't why aneurysms sometimes occur, but why they don't normally happen. We have to ask why an arterial wall, as extensible, noncreeping, and resilient as any good rubber, nevertheless behaves in a much friendlier manner. What we must consider an answer to the question is very much a matter of the explanatory level at which we seek it. At a fairly superficial level we already have an answer—arteries don't behave like balloons. Pump one up with a heartbeat, and it expands all over. If you decide that such an answer is unsatisfyingly glib, then you're on the reductionist path that scientists find both powerful and seductive. You ask, then, about the mechanical properties of the walls of arteries, to see how they might be responsible for the odd behavior. After stretching samples of arterial wall with some procrustean testing machine, you find mechanical behavior appropriate to explain the behavior of arteries. Then at some meeting, someone punctures your balloon by asking about the structural basis of that mechanical behavior, and you're off to look at the arrangements of the materials that combine to form arterial wall. That, naturally, simply cries out for an investigation of the mechanical behavior of those several materials, which, in turn, raises the issue of the chemical bases of such mechanical behavior. If there's an end to the reductionist process, it's an arbitrary or at least a pragmatic one. The level to which an investigation penetrates turns on a lot of mundane factors as commonly as it does on some direct scientific issue—factors such as the personality and background of the investigator and the particular technology with which that investigator is comfortable.

Digression aside, arteries should surely stretch strangely. Say we stretch a sample of material by pulling with known increments of force and then measure the resulting extension of the material—we get data that can be displayed as a graph of force versus extension. A little manipulation of the numbers can eliminate any particularity due to the specific sample, so we can plot some numbers on a graph to describe the material itself. We divide the force we applied by the cross-sectional area of the sample to get what's called *stress;* we divide the length we measured as the sample extends by the unstretched length to get what's called *strain*.[3] A plot of stress versus strain is thus a kind of generalized force versus extension graph, as shown in Figure 7.2.

Many common materials give something approximating a straight, upwardly sloping line on a stress-strain graph. Roughly, at least, a rubber band behaves that way; a steel wire gives an even better approximation of that simple situation. (That straight line, by the way, is why, with steel springs, scales for weighing can have uniform increments marked on their dials.) Such materials are called *Hookean*, after Robert Hooke (1635–1703). For each unit of force added, the material extends a proportional unit. The material may be of any degree of stiffness, but

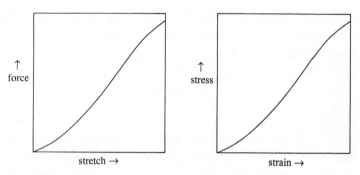

Figure 7.2. The graphs are similar, but the one on the left is specific to a specimen, showing the relation between applied force and the resulting stretch. The one on the right applies to the kind of material rather than the particular specimen; it uses force per area (stress) and extension relative to original length (strain).

that direct proportionality between force and extension still holds. Hang several equal weights one by one from a hook on a rubber band (fishing sinkers will do) and you ought to see a fairly Hookean result. Rubber usually gets a little stiffer just before it breaks (mentioned earlier in connection with balloons), which gives its graph an upward hook on the upper end.

Such upward-sloping straight lines aren't what one gets from a test of arterial wall. Arterial wall gives a very different sort of line (Figure 7.3), one that has an ever-increasing, rather than constant, slope. That means that a disproportionate force is needed for each incremental bit of stretch—the thing gets stiffer as it stretches further, which is the basis of the special trick of the arterial system. As the vessels expand, pressure inside is increasingly effective at generating tension in their walls—that's the unavoidable consequence of Laplace's law. *But that tension, the stress in the walls, is decreasingly effective in causing the walls to stretch.* It all comes down to that curved, J-shaped line on the stress–strain graph, which means no aneurysm.

How to Make an Arterial Wall

Moving further along on the reductionist path, we ask another question. What is it about the structure and materials that form an arterial

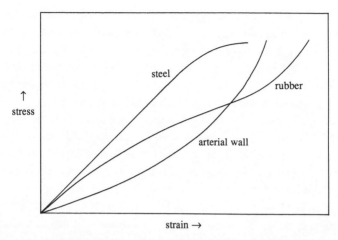

Figure 7.3. Stress–strain graphs for rubber, steel, and arterial wall. To emphasize the differences in the shapes of the curves, I've used arbitrary and nonequivalent scales on the axes.

wall that produces the peculiar but crucial increase in stiffness as the wall is stretched? At this point it's important to recognise that arterial wall isn't really a material in the same sense as are rubber and steel—it's better described as a structure. Or, depending on how you choose to define things, it's a composite rather than a simple material made of a single chemical component. What matters isn't just composition but the arrangement of its constituents. From our present mechanical perspective, the important elements are two fibrous proteins of very different properties. And not only are their properties different, but they are arranged in quite distinctive fashions.

First, there is *elastin*. As the name implies it's highly extensible stuff. To get some feeling for its character, it's handy to have a large and unadulterated sample. The nearest thing to that ideal is a ligament that runs along the top of the neck of many grazing mammals; this *nuchal ligament* is a yellowish rope that can be dissected out of an unsliced lamb neck. (Try it if you are so disposed). Elastin can be stretched quite far before breaking—to a little over double its unstretched length. And not only is it highly extensible, it's decently resilient: about three quarters of the energy of extension is given back when the substance is allowed to return to its original length.

Then there's collagen, which we met before in connection with hearts, tendons, and tough meat. Collagen is much stiffer than elastin, about 1000 times so. It's also less extensible,[4] stretching by only about a tenth of its initial length before suffering irreversible structural change. But it's even more resilient, returning over 90 percent of the energy of stretch. It's so good at storage and release of energy that it was used as the basic elastic of the most powerful pieces of ancient artillery, the ballistae of the Greeks and Romans. (These were far more efficient machines than any medieval catapults.) Throwing a 90 pound rock 400 yards with a bunch of twisted cow tendons is surely impressive; but it really is a shame that we humans seem always to put our best efforts into military, ritual, or funereal technology.

While arterial wall has both kinds of fibers, they're arranged, as already noted, in quite different ways. The elastin fibers are largely straight when an artery is unstretched, so increasing the pressure inside stretches the fibers. They resist, of course, but because they're simply not very stiff they don't resist strongly. By contrast, the collagen fibers in an unstretched artery are normally kinked and bent. As pressure goes up and the wall stretches, an increasing number of these much stiffer fibers reach their normal resting lengths, after which they're properly stretched by further increases in pressure. So the wall gets stiffer and stiffer as it's stretched by the tensile stress of an increasing pressure inside. This contrasting arrangement of two very different kinds of fibers proves entirely adequate to explain the J-shape of the curve of the graph of stress against strain. Either component alone follows Hooke's law at least roughly, but the combination does something quite different. Over the range of pressures to which arterial wall is subjected, the stiffness changes no less than fiftyfold. Figure 7.4 is a somewhat diagrammatic representation of the phenomenon—the actual microanatomy is somewhat more complex, for reasons of no immediate relevance to the present story.

We've turned to the cephalopod mollusks at several points to see how a different lineage of complex and active animals goes about a task. A squid or octopus ought to have functional constraints on its circulatory pipes similar to those of a vertebrate since both use pulsatile hearts to propel blood at substantial pressures. The pipes ought to be stretchy, and with Laplace's law unavoidable, they shouldn't have walls that follow Hooke's law. About 10 years ago, John Gosline and Robert Shad-

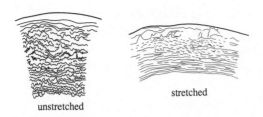

stretched

unstretched

Figure 7.4. How collagen fibers are arranged in arterial wall to produce the peculiarly J-shaped stress-strain graph of Figure 7.3.

wick took a look at the aortas of members of a large species of octopus. As they pointed out, elastin is exclusively vertebrate stuff, and its appearance coincides with the arrival of the vertebrate sort of advanced circulatory system. They found that the octopus has a different but functionally quite analogous highly extensible protein in its aortic wall, one previously quite unknown. This "octopus arterial elastomer" turns out to be the main player in the arrangement to increase stiffness as the wall is distended, with collagen coming into play only at the highest pressures. (Collagen, as mentioned a few chapters back, is ubiquitous among animals.) Despite the difference in that detail, the shapes of the stress–strain curves they obtained for octopus aortic wall were barely distinguishable from those of mammals.

More recent work has shown that the same behavior, undoubtedly responsive to Laplace's law, characterizes the blood vessels of crustaceans, animals quite distinct in lineage from either cephalopods or vertebrates. Their major arteries are similarly elastic, with that increase in stiffness as distension is increased. They also have the same high resilience, giving efficient smoothing of the pulsatile flow from the heart. Crustacean extensible fibers are made of still another protein; again collagen is the less extensible component.

The convergence is a deeply satisfying one to the student of biological function. Our notion of the functional significance of the presence and arrangement of two different elastomers in arterial wall tacitly predicts that an octopus or a crab should do something analogous to what a vertebrate does. Verifying that prediction not only tells us a little about octopus and crab, but it does several other things as well. First, it suggests that similarly elastic blood vessels will be found in many animals not yet specifically scrutinized. Second, it lends credence to our functional interpretation of the arrangements of our own arterial walls. On a more general level, it once again illustrates the explanatory power of a comparative approach to figuring out the principles of animal function.

So What About Veins?

Laplace's law applies to all cylindrical conduits while Hooke's law provides an idealized frame of reference for extensible materials. Stress, strain, stiffness, resilience, and extensibility are either measured variables (stress and strain) or properties of materials (stiffness, resilience, and extensibility). So while all the applications have so far been arterial, the underlying physics and geometry is much more general. We turn now to veins, a set of pipes as different from arteries as your drain pipes are from those of your water supply. Despite the common primary function of carrying blood from place to place, arterial and venous systems operate under substantially different local conditions and have quite distinct secondary functions.

A set of arteries, dealing with the output of a pulsatile pump, works as a pressure buffer, alternately stretching and shrinking, storing and releasing energy, and thus smoothing variations in both pressure and flow. All this is done at rather high levels of pressure. By contrast, a venous system operates at low pressures and receives a nearly steady flow from the capillaries and other elements of the so-called microcirculation. If you cut an artery the blood comes out in spurts corresponding to the ventricular contractions. By contrast, if you cut a vein the blood flows out quite steadily. In either case, do something about the situation, but remember that you have well over a gallon of blood. Loss of an ounce or two may make a terrible mess, but it won't otherwise be noticed.

This matter of losing blood, though, gets us right to the main secondary role of the veins. If you're relieved of a pint or so of blood, you're not seriously disabled by the deletion; you can be quite sanguine about the exsanguination. Your blood pressure changes very little, which means that your heart and arteries are inflated very nearly to the same amount as they were previously. It turns out that the volume change has been accommodated almost entirely by the veins. The arteries are your pressure buffer; the veins are your volume buffer. At any instant about 80 percent of the blood moving through the systemic circulation is in the veins. Thus even if a change in blood volume is reflected entirely in a change in venous volume, the latter needn't alter disproportionately.

It's no mean trick to maintain normal function in a closed system of pipes in the face of variations in total volume, and it demands that venous pipes have properties quite as specialized as those of arteries. They have to be larger, overall, than arteries, as noted already. They have to be flabbier, since volume changes will occur in the context of very much lower pressures. (You might happen to recall from Chapter 4 that venous pressure runs downward from about 15 mm Hg, an order of magnitude lower than arterial pressures.) They are indeed flabbier— for a given vessel diameter, a vein has walls only about a third as thick as those of an artery.

Arteries swell when the pressure inside rises, and so do veins. Expansion is accommodated by much the same scheme, using fibers of collagen and elastin in concert. For flabby pipes, veins are surprisingly strong, capable of withstanding up to around five atmospheres (nearly 4000 mm Hg) before rupture.[5] Arteries shrink again when pressure drops; veins do this and more—they become noncircular in cross section. Quite literally, they collapse. But the collapse isn't something that happens when pressure drops below some critical value. Instead, it's a gradual shift in cross section from circular to elliptical to roughly a dumbbell. *Collapse* strikes me as an unnecessarily pejorative word for a gradual and functionally useful change in shape, but that's the usual term. In all, the change in cross-sectional area and thus of volume inside is far greater in veins than in arteries. Of course, that's just the requirement for elements that offset volume changes in the system as a whole.

To me, this arrangement to accommodate volume changes is only the second most clever aspect of the design of the venous system. The real fun comes in when we realize that venous pressures are very much lower than the ordinary hydrostatic pressures in upright animals 5 or 6 feet tall. Venous pressures of 10 or 15 mm Hg have to be viewed in the context of, say, a gravitationally induced pressure of 90 mm Hg in a vein on the top of one's foot when one is erect but at rest. (That's the consequence of a column of blood 4 feet high.) How can the blood find sufficient incentive to climb up again to the heart? To complicate things further, we persist in changing our postures suddenly and unexpectedly. As was the subject of a lot of talk in Chapter 4, we are a bunch of hydrostatically ordinary manometers, whatever our other pretensions.

Veins are commonly located closer to the skin than are arteries, and we can take advantage of that to illustrate our hydrostatic mundanity with a simple maneuver. Here's how to estimate the pressure in your right atrium with nothing fancier than a metric tape measure. Sit upright but relaxed in a chair, letting one arm dangle until the veins are well-filled and visibly bulging with blood. (This works less well in tight-skinned youngsters.) Then slowly raise the arm until the veins in the back of the hand just begin to collapse. The height above heart level at which this occurs gives the pressure, in millimeters of blood, in the right atrium. Divide by 13.6 and you have the result in millimeters of mercury. What you've done by raising the arm is to establish a hydrostatic head of pressure just sufficient to force blood into the atrium, so blood runs out of the hand without any distention of the walls of the veins.

Now we'll go back to that figure of 90 mm Hg in a foot vein. Start walking and the pressure drops to about 20 mm Hg, which seems a most peculiar behavior for a manometer ("personometer" or "pedometer," perhaps?). What your legs have is an auxiliary pumping system. Think back to the discussion in Chapter 2 about the minimal requirements for a pulsatile pump—it took only a squeezer and two valves to do the job. Veins are easy to squeeze, muscles are there to squeeze them, and that's exactly what happens in your legs when you walk. The necessary valves are distributed through the veins, as discovered by Fabricius and correctly interpreted by William Harvey (Chapter 3). They're simple but effective sets of cusps, as shown in Figure 7.5, that ensure that any increase in pressure attending muscle contraction will propel venous blood heartward. They're particularly prevalent in the big veins of the extremities, while they're absent in the largest veins near the heart and in the smallest veins elsewhere. Giraffes, whose arterial systems were mentioned earlier, have a particularly fine set of

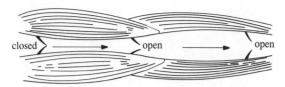

Figure 7.5. The functional arrangement of the venous valves of your appendages.

valves in the veins of their necks, almost certainly to keep blood flowing properly when they bend down to drink.

Of course, the valves are of only minor help, if any, without muscle activity. If you insist on remaining erect or even sitting straight without "working" your legs, the pressure increases in them, despite the valves. That increase has several consequences. The least of these is that higher pressure increases leakage of fluid out of the capillaries and into the spaces among the cells. So your feet swell as fluid accumulates. We seem to have some neural circuitry that gets tickled by the problem, as you can notice if you try to sit on something from which your legs dangle. As children find out, it takes concentration to keep the legs from moving. The muscles "want" to work, and only with attention can extensors and flexors be sufficiently synchronized that the legs remain stationary. If you're confined for a long period, for example in a typical tourist-class airplane seat, your legs and feet swell, and you have to loosen your shoes a little to get them back on. I won't take a long trip any more without wearing heavy support stockings, which seem to reduce the post facto discomfort. (Incidentally, alcohol is both a muscle relaxant and a vasodilator, so it exacerbates the pooling of blood in feet and legs—even a little drinking while flying is likely to make a person dizzy.)

Fluid pooling from leg muscle inactivity when you're upright can lead to more serious trouble. Retention of fluid reduces circulating volume and blood pressure, and that in turn can make a person dizzy or even lose consciousness. I'm told that it's especially dangerous to be caught in a crevice in, say, a cave. If legs go inactive, you can faint; conversely, if you pass out your legs become flaccid. Either way there's a fair likelihood that you won't ever wake up since the brain is then

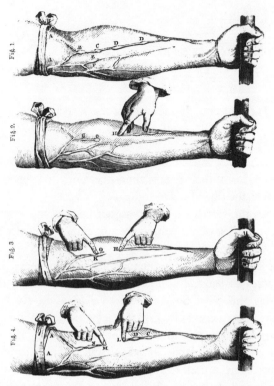

Figure 7.6. Harvey's illustration, showing how one can demonstrate the action of the venous valves. He advises using a laborer or someone with conspicuous veins as subject. The version here is the redrawing found in Willis's English translation of Harvey, done in 1847. It's a little clearer than the original and thus more useful for present purposes.

starved of oxygen. It has been said that this fainting without being able to fall over is the immediate cause of death in crucifixion—neither nails nor exposure are anywhere nearly as hazardous. Of course, the practical bit of advice is to keep the head of an unconscious person as low as possible—death from anoxic legs isn't a serious hazard.

William Harvey gave instructions for demonstrating the presence and action of venous valves that can still serve, since we obviously haven't done a lot of evolving in the intervening fifteen generations. You might try the procedure on yourself, although a little less dexterity is required if a second person does the fingerwork. Figure 7.6, from Harvey (the only one in his book), illustrates what has to be done.

[L]et an arm be tied up above the elbow as if for phlebotomy (A, A, fig. 1) [As explained elsewhere, he meant a light tourniquet so arterial blood can get into the arm but venous blood can't escape, thus inflating the veins, as we do before drawing blood]. At intervals in the course of the veins, especially in laborers, certain knots or elevations (B, C, D, E, F) will be noticed, not only at the points of branching (E, F), but also where no branch enters (C, D): these knots or risings are all formed by valves, which are thus reflected externally . . .

Moreover, with the arm still bound and the veins full and distended, if you press at one place in the course of a vein with a fingertip (L, fig. 4), and then with another finger push the blood upwards [toward the heart] past the next valve (N), you will see that this part of the vein stays empty (L N), and that the blood cannot flow back.

One cannot easily contrive a better description or demonstration!

Notes

1. My colleague, Stephen Wainwright, says that if you can't show it with balloons, it's probably not worth knowing.

2. Matthew Healy reminded me of the citation. One cannot easily anticipate all the benefits of harboring graduate students.

3. This kind of procedure is a common way to treat experimental data. Again, we're usually interested in properties of a material, not of a particular sample of the material. It's no accident that we divide by three distances, two hidden in the cross-sectional area (an area represents the product of two distances) and one in the original length—those fully adjust for the three dimensions of a solid specimen.

4. Someone may notice that I'm avoiding using the words "elastic" and "stretchy" to characterize materials such as these fibrous proteins. The trouble is that in common usage, they confuse the properties of stiffness and extensibility. Stiffness is the stress needed to get a given strain, while extensibility is how far, relative to original length, something can be stretched before getting into some kind of mechanical difficulty.

5. Thus no problem in bypass surgery results from using veins grafted around occluded coronary arteries.

8 *Hearts, Again*

Hearts have repeatedly appeared in this account, and the reader may be wondering why the writer doesn't get them off his chest once and for all. This circulatory machinery, though, is a well-integrated and coherent system, so we commonly need information about one part to take a more penetrating view of another. In addition (and more specifically) we're now armed with the notions of diffusion, continuity, Laplace's and Hooke's laws, the Reynolds and Hagen–Poiseuille relationships, and so forth, which we weren't just a few chapters back. In a proper textbook the nonbiological underpinnings would be done at the start; but a textbook typically talks to a relatively involuntary reader who must put up with such deferral of gratification. (Perhaps the main difference between a nontext and a text, though, is that the nontext is designed to appeal to people looking for a decent level of novelty while the text is selected by people who often prefer a minimum of novelty.)

Macropumps and Micropumps

The phenomenon of viscosity made a pump mandatory. Two kinds of biological pumps have come up so far, valve-and-chamber pumps and peristaltic pumps. The valve-and-chamber scheme characterizes most hearts as well as the auxiliary pump made up of muscles, veins, and venous valves. A few peristaltic hearts occur here and there in the animal kingdom: the reversing hearts of tunicates (sea squirts) were mentioned, as was a giant worm that had some peristaltic blood vessels. Peristaltic peripheral vessels in cephalopods are softly rumored. What hasn't been brought up is an entirely different possibility for propelling blood through pipes. Maybe the best way to introduce the matter is to remind ourselves of the diverse sizes of the elements of circulatory systems—at least a 10,000-fold range of diameters from any capillary to a whale's aorta. All of these systems use both large and small pipes, with high-speed flow in the large ones and low-speed flow in the small ones. So far we've tacitly assumed that the natural location for a pump is in the part of the system with the large pipes and the fast flow.

Taking a sufficiently broad view, both extremes must constitute possible locations for some form of pump. We might call pumps located where flows are fast and total cross sections small "macropumps." These would then include all of the hearts, venous auxiliaries, and peristaltic pumps so far mentioned. Obviously the only reason to invent such a name is to distinguish them from things to be designated "micropumps." The latter, of course, are the opposite—pumps located in the smallest vessels, where flow is slowest and total cross section greatest. The idea of a circulation driven by micropumps is certainly a strange one. We'd have the pumping system distributed through the microcirculatory vessels, the vena cava could be directly connected to the pulmonary trunk, and the pulmonary vein to the aorta. One might write some bit of science fiction in which the aliens are distinguishable only by the absence of any heartbeat. In ourselves, of course, capillaries and arterioles do no pumping—why introduce this heartless straw man?

As it happens, we have gotten into another of those "what if it were otherwise" questions about which so much was made in Chapter 6. Just as we could recognize organisms that did all of their material transport by diffusion, we can similarly find organisms that use micropumps rather than macropumps. These micropumps are usually not parts of what we'd ordinarily call circulatory systems, but they're nonetheless involved in moving liquid around. Again, there's some insight about our own arrangements to be obtained through a look at sharply contrasting ones.

In some of these cases the volume flow rates achieved by micropumps are far from trivial. A clam or other bivalve mollusk has enormous gills that take up a large fraction of the space within its shell (Figure 8.1a). Those gills are covered with cilia, tiny filamentous organelles that protrude from cell surfaces. By waving in synchrony, they pump water into the clam, across the gills, and out again. In a class I once taught at a marine lab, one of the students, Edgar Meyhöfer, found that a scallop or mussel moved a volume of water equal to its own volume *every 3 or 4 seconds*. These gills are only secondarily respiratory structures; such bivalve mollusks mainly process such large quantities of water for feeding. Even clams need food, and they get it by separating water from the tiny organisms and other edible particles suspended in it.

Sponges (Figure 8.1b), commonly regarded as the most primitive of multicellular animals and which got notice earlier in connection with Murray's law, feed in a similar fashion. Water is drawn in through tiny pores, and it passes into pumping chambers lined with flagellated cells. These flagella pump water by waving around in a way similar to the cilia of clams (although they don't beat in such tidy synchrony). The water then goes through another set of pipes that get bigger and bigger and of lower and lower aggregate cross-sectional area, until the water is finally ejected back into the ocean. Again the volume flow rates aren't low. Henry Reiswig and I independently have found that active sponges in a warm ocean pump an amount equal to their own volume *every 5 seconds*.

That seems to be the general picture for such suspension feeders—an awful lot of ocean has to be processed to get enough to keep body and soul together.[1] An ounce of food from a ton of water, someone once calculated. Some use can be made of the natural movements of water in nature to push it through a filter. Sponges thus get some help from water currents, and many tiny insect larvae in streams are entirely dependent on stream flow. Really large-scale pumping isn't at all uncommon—clams and sponges aren't rare or obscure, nor are they the only big-time pumpers. To put those pumping rates in some context, they're about 100 times as high as that of the heart of a moderately active human, where the rates are expressed relative to body volume. When it comes to moving liquids, active suspension feeders are simply in a league by themselves. Even the actual speeds of flow are fairly high. A sponge ejects water at about 20 centimeters (8 inches) per sec-

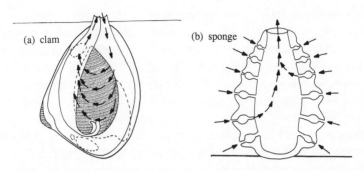

Figure 8.1. The way micropumps are used in (a) clams and (b) sponges.

ond, roughly the same as the speed of flow in our aortas when we're relaxed.

A century or so ago biologists wondered how the activities of individual, uncoordinated cells could push anything at such high speeds—getting more than 1 millimeter per second with a cell-level pump seemed quite out of the question. What they didn't appreciate was the principle of continuity. This detachment from physical science and the practical world has bedeviled biology and, in many quarters, continues to do so. The notion of continuity[2] is nothing new—it was quite clearly appreciated by the people who designed the great sewer system (the "Cloaca Maxima") of ancient Rome. The point for us, not to get too far afield, is that the use of micropumps doesn't at all preclude achieving quite high speeds of flow. The pump needn't directly produce the high speeds nor even be near the fastest flows. All that's needed is a collecting manifold, a set of pipes of decreasing total cross-sectional area downstream from the pump, or the opposite, a distributional manifold somewhere upstream. So it isn't so preposterous to imagine a circulatory system in which the capillaries are lined with cilia and a heart is quite superfluous.

In one respect, pumping with cilia-lined pipes has an advantage over using valve-and-chamber pumps. Recall the discussion of the way speeds of flow varied across a pipe (Chapter 5)—flow was normally absent at the walls, maximum in the center, and formed a parabola if drawn as a graph of speed against position across the pipe. Then recall our talk about diffusion in Chapter 6. If the two phenomena, parabolic flow and diffusion, are put together, one has a Bad Thing. With parabolic flow, a disproportionate fraction of the fluid flows near the middle of the pipe and concomitantly little flows near the walls, clearly bad for a process as disadvantaged by distance as is diffusion. Thus if material is being diffusively exchanged across the walls, as in capillaries, then the normal is a kind of worst case.

The advantage of a ciliary or wall pump becomes apparent if we consider several possible ways flow might vary across a pipe, putting a number that gives some indication of performance on each. Consider, then, what we might call "effective diffusion path length"—how far from the pipe wall the average bit of fluid travels as it moves down the length of the pipe. For normal, parabolic flow, that path length is a little more than half the radius of the pipe, which means that the average bit of fluid moves along a little nearer to the pipe's axis than to the wall (Figure 8.2b). What if the speed of flow were uniform across a pipe, that is, with the same speed everywhere from center to wall? (Again we're taking a "what if it were otherwise" approach to revelation.) In that case the average bit of fluid would travel at a third of the way from the wall to the axis of the pipe. In other words, for uniform ("plug") flow the effective diffusion path length would be a third of the radius of the pipe (Figure 8.2a). Decreasing the distance from over a half to just a third is no small advantage—recall from Chapter 6 how drastically diffusion depended on distance.

To complete the contrast, consider what the effective diffusion path length would be if the pump were located on the walls, as a thin layer of cilia would be, and the resistance to flow were some nonciliated pipes elsewhere in the system. The conditions for derivation of the Hagen–Poiseuille relationship have now been reversed. Maximum speed of flow occurs right near the walls, and minimum speed happens in the middle (Figure 8.2c). The exact effective diffusion path length depends on the conditions one chooses for the calculation; if we pick the case where flow on the axis is just zero, then that path length comes out to a fifth of the radius of the pipe. That's much better even than the situation for uniform flow, and far better than that for ordinary parabolic flow. Putting the pump on the wall makes the fluid flow nearer the wall.

Even we ourselves have pipes lined with cilia. The mucus lining our tracheal tubes is moved by such cilia, and the uterine or Fallopian tubes

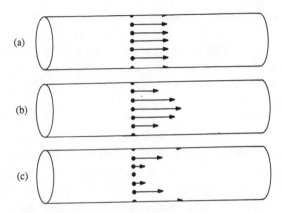

Figure 8.2. Several ways flow speeds can vary across a pipe. In (a) flow is uniform across the pipe; (b) shows the result for normal, parabolic flow; (c) presumes that flow is zero in the center and inversely parabolic overall. Again arrow length reflects flow speed.

are ciliated as well. Our capillaries, however, are unciliated, and we do have hearts. What advantage might there be to using a central, macro-pumping heart (obviously the people's choice among blood circulating arrangements)? Michael LaBarbera and I have argued that the advantage is an economic one. Circulating blood isn't any staggering tax on our metabolic expenditure—it's only around 10 percent of resting metabolic rate. Still, 10 percent isn't trivial either, and ciliary pumps (I won't go into the evidence here) just aren't very efficient energetically, whatever their other virtues.

The economic argument, then, demands some special rationale for any use of ciliary or flagellar pumps. But that's not hard to provide—they seem to be used where some special benefit is obtained by their use or where other special conditions apply. For the suspension feeders, the cilia or flagella (really much the same thing for our purposes) are intrinsic components of the filtration device itself, the machinery with which mixer and munchies are made to part company. For tracheal or uterine tubes, not only is flow slow, but sheets of mucus are moved. Mucuses aren't liquids in quite the normal sense, so different arguments and rules will apply.

I think it highly likely that, looking globally, more liquid is pumped through living systems by micropumps than by macropumps. Suspension feeders contribute a lot, but they're probably not the main users. The really major case is the pump that draws most of the water up trees. The pump is an evaporative one—water changes from liquid to vapor and goes off into the atmosphere. By a clever bit of design, air cannot enter and fill the tree's pipes. Thus as water changes from liquid to gas, more liquid water (inextensible as well as incompressible stuff) is raised from the roots. Columns of water extend in thin pipes up the periphery of trees; they branch into the branches; they penetrate the petioles; and they end, as unbroken liquid streams, in the walls of cells within the leaves.

The area of exposure to the atmosphere at these cell walls is enormous—for a small orange tree with 2,000 leaves, an aggregate area of an acre and a half was once calculated. So (by continuity again) even a slow rate of evaporation over that huge area can draw water up the pipes around the periphery of the trunk at substantial rates. A centimeter per second, which we are wont to regard as fast for any activity of a plant, isn't uncommon. The overall amounts of water lifted for a given area of ground over a growing season are truly enormous—for an acre of bamboo forest in Java, the rate has been estimated as 1.5 million gallons each year. For rain forests in general, more than half the rainfall is commonly returned to the atmosphere by this process, and it has substantial climatic effects.

What, then, about the economic counterargument? It just doesn't apply to this so-called transpirative process in plants. Transpiration is directly driven by solar energy with no metabolic broker. Left to itself water will evaporate, but evaporation takes energy, which is why one can be cooled by sweating—you supply thermal energy to drive evaporation. Leaves may be cooled by transpiration, but in anything but the shortest of terms, the source of energy is the sun. The energy is simply there for the taking by organisms that have to be exposed to sunlight to manufacture food. Metabolic cost is simply not a factor, and thus micropumps hold sway in trees and most other terrestrial plants.

But for large animals—as put in Psalms, "he fashioneth their hearts alike"; and he "maketh glad the heart."

Regulating the Heart

As John Donne might have said to William Harvey (more likely than not they knew each other in the small London world of the early 1600s) no ventricle is an island, entire of itself. The two sides of the heart of every bird or mammal must pump precisely the same amount of blood measured over any time interval greater than a few seconds. They must also maintain that equal output despite enormous variation in the demands on the heart as a whole, depending on what an animal is up to. Keeping the two sides balanced and adjusting output in response to demand both point a finger at regulatory mechanisms, of which hearts have a fair hierarchy.

For convenience and by convention we consider "cardiac output" to be the output of one side only, just keeping somewhere in the back of the mind that the real output is exactly twice that. The convenience will be evident when we talk about oxygen transport since conventional cardiac output times oxygen concentration gives that rate at which oxygen is supplied to the (nonpulmonary) cells of the body. The output of a half-heart represents the product of two variables, how much blood is ejected in each stroke and the frequency of the heartbeat. We thus multiply two factors, stroke rate or frequency and stroke volume, to get volume per unit time. Stroke volume, how much blood is ejected, is the difference between maximum and minimum ventricular volumes, called diastolic (relaxed) and systolic (contracted) volumes, respectively.[3] In short, cardiac output is pulse rate times the difference between greatest and least left-ventricular volumes.

First, then, we have the problem of equalizing the outputs of the two sides of the heart. Heart rate, of course, is no problem at all—it cannot differ between the pumps in anything this side of an octopus. Or we might view heart rate as a problem of omission, since equalization of volume flow rate cannot be done by tinkering with the rate of beating. Only stroke volume remains as a possible variable quantity since no anatomical constraint prevents at least minor adjustments of the extremes of volume of the two sides. The equalization scheme proves to be a very simple one, based on the intrinsic properties of the pump and ultimately on the general behavior of muscle, rather than depending on special machinery. Within limits, the tensile force a contracting muscle can develop depends on how far it has already shortened. To put it another way, if a muscle pulls on a substantial load, the force the muscle can develop varies with its length. A muscle stimulated to pull while at its resting length can pull harder than one that is stimulated when it has already shortened by 10 or 20 percent. Now imagine that a little extra blood gets into one ventricle, never mind how. The ventricle is then larger during its diastolic phase, the muscle fibers are initially longer, and they can therefore give a stronger contraction, pushing harder on the blood. In addition, of course, with a larger diastolic volume the overall stroke volume can be greater. Thus a little extra input automatically calls forth increases in both pressure and volume in the output.

This adds up to a nicely self-compensating system. If more blood flows into either side (or both sides) of the heart, it doesn't accumulate, but instead stimulates that side (or both) to work harder and thus to push the extra out. Recognition of this automatic mechanism grows out of the work of the nineteenth-century German physiologist, Otto Frank (mentioned earlier in connection with the windkessel model), but it was elaborated in its present form by a Briton, Ernest Starling, and is most commonly known as *Starling's law of the heart*. However, the original idea that this was the principal control over cardiac output has not held up well. Still, it does seem to be the bottom line when it comes to the basic problem of synchronizing the outputs of the two sides. Perhaps we can be forgiven a little biological hubris—almost no pump in common technological use has this nice behavior. Two pumps in series typically need some special device, such as a small bleeder shunt around one of them, to work efficiently together.[4]

The Starling kind of compensation may take advantage of some very basic biology, but it eventually runs afoul of some even more basic physics or geometry. The ghost of Monsieur Laplace lurks in the walls of all pressurized vessels, which the heart most certainly is. The heart muscle produces tension by contraction, and that in turn produces the pressure that squeezes blood outward. If the heart gets bigger, by excessive refilling, then a given level of tension will produce a *lower* pressure. Tension, remember, is proportional to pressure times radius—at a constant tension, an increase in volume or girth causes a decrease in the pressure within. We used the law for cylindrical blood vessels, but it applies quite as unavoidably to spherical, conical, or elliptical chambers. If you have trouble envisioning what's happening, consider the extreme case. Imagine a completely flat, muscular wall, the case if the radius of curvature were infinite. As the radius or girth approaches infinity, the pressure must approach zero, so no matter what the tension of the muscle, if it's part of a flat wall it can't generate any pressure inside. That's intuitively reasonable, since a wall has to be wrapped around a chamber for a contraction of the wall to increase the pressure inside the chamber. As Alan Burton put it for the benefit of medical students,

> Thus the larger the volume in diastole
> The greater the output was likely to be.
> But when the heart reaches a much larger size,
> This leads to Heart Failure, and often, Demise.
> The relevant law is not Starling's alas,
> But the classical law of Lecompte de Laplace.

As earlier, the Laplacean dilemma can be illustrated with balloons. If two similar balloons are attached to the ends of a Y-shaped tube, as in Figure 8.3, then the pressures inside the two must be the same. One might expect that they'd inflate to similar volumes when one blows into the lower arm of the Y. Not so—one balloon usually inflates almost to bursting before the other so much as gets started. Perhaps one balloon is much weaker. Again, not so—close off the long arm of the "Y," cup your hands around the larger balloon to squeeze it, and thus inflate the smaller. Beyond a certain volume, you don't even have to squeeze the first balloon—the second will quite spontaneously complete its inflation. Once again we see that pressure is more effective in generating wall tension in whichever balloon is larger. A given level of wall tension is *less* effective in generating internal pressure in the larger balloon.

Normally there's no problem, since normal ventricular wall doesn't behave any more like a simple rubber membrane than does normal arterial or venous wall. It has a similarly curved graph of stress against strain, and so it's elastically stable. The trouble comes when part of the wall of the heart is weakened, perhaps as a consequence of an *infarct* (what's commonly called a "heart attack"). If that weak part bulges outward, it's called a *cardiac aneurysm*. Bogen and McMahon[5] looked

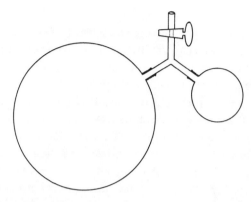

Figure 8.3. Inflating two balloons on a Y-tube. Only one expands appreciably.

at the consequences of such bulging. Were the bulge made of normal heart wall, there'd be no trouble unless the bulge were larger (had a larger radius of curvature) than the rest of the chamber, but an aneurysm has a wall that's thinner and weaker than elsewhere. If it's sufficiently weak and the aneurysm sufficiently large, then blow-out is an alarming possibility. Otherwise, the system tends to return spontaneously to sphericity since a little bulge has a lower radius of curvature than elsewhere and, other things being equal, resists further expansion more forcefully.

(That, incidentally, is why as it's inflated a spherical balloon stays spherical even if its walls have not been manufactured to perfect uniformity. It's possible you've never worried about the problem of maintaining near-sphericity in inflating balloons. It's not just a good idea, it's the law.)

Enough of such depressing pathology. It's been mentioned already that the demand on the heart varies a lot depending on what an animal happens to be doing. A human heart pumps about 1.5 gallons each minute at rest, but in vigorous activity in a person of good aerobic fitness that output may go up fivefold. Through various signals, to which we'll return in a later chapter, the operation of the heart is pretty drastically altered in the process. There are two main variables available for adjustment, stroke volume and the rate of heartbeat. Interestingly, stroke volume changes only a little, going up from about 70 cubic centimeters (2.4 ounces) to about 100 cubic centimeters (3.4 ounces). The heartbeat, though, gets much more frequent, rising from perhaps 60 per minute to 180 or even more. So we do most of our compensation for activity by changing rate rather than volume.

Beyond this basic picture a few details might be mentioned. Heart rates above 180 are neither uncommon nor signs of trouble, especially in young people. However they don't move a lot more blood since at very high rates the stroke volume decreases as the rate increases, and one change offsets the other. Blood pressure increases as the heart beats faster, quite a reasonable thing if one pushes blood around faster since higher speeds mean more viscous friction in the vessels. At the highest rates, though, pressure drops off a bit. The force a muscle, any muscle, can generate drops as the rate of contraction increases; at high heart rates, heart muscle gets pushed up to high speeds of shortening. When I'm tested on a treadmill, running at increasingly sadistic speeds and grades, my blood pressure rises steadily; but it drops a little shortly before I cry "hold, enough" to tell that day's MacDuff to terminate the test.

Several comparative notes might be mentioned. First, it's clear that use of stroke rate rather than volume for adjustment of output isn't some absolutely necessary precondition of the valve-and-chamber kind of pump. Fish mainly do the opposite, making little change in heart rate but major adjustment of stroke volume as they swim faster. (I have a notion of why the arrangement makes particularly good sense for a fish that I'll get to shortly.)

Second, one might suspect that especially high heart rates raise difficulties for muscle action, for timing of the various events, and so forth. It's therefore worth looking at the highest of all mammalian rates, which, naturally, occur in the smallest ones, shrews. Shrews turn out to be big-hearted creatures, at least relative to their minute sizes. Thus members of a species with an average body weight of 7.6 grams (a quarter of an ounce—shrews, of which these aren't nearly the smallest, make mice look like men) have hearts of about a percent of body weight, nearly twice the norm for mammals. Heartbeat rates, however, are lower than expected even for those larger hearts. The resting rate predicted by extrapolating from data on larger mammals is 817 per minute; the actual rate is only 690—still, over eleven per second. Shrew hearts are unusual in shape as well as size. Normal ventricles are about as wide or even a little wider than they are long; shrew ventricles are twice as long as wide. It has been argued that Laplace's law gives an advantage to a skinnier heart by permitting higher pressures and greater efficiency, at least when the creature is small enough so other considerations don't preclude such a shape. The idea certainly sounds reasonable enough.

Bird hearts are generally somewhat larger than the hearts of mammals of the same body size, but for a given size of heart, a bird heart beats a little more slowly. Thus overall cardiac output is similar. Hummingbirds have large hearts, even by avian standards, just as shrews are exceptional among mammals. The highest heart rates known are for active hummingbirds and shrews; interestingly the values, 1200–1300 per minute, are about the same in the two—over twenty beats per second!

Refilling a Ventricle

In a piston pump, one stroke of the piston ordinarily expels fluid and the other stroke draws in new fluid. A set of valves determines that the intake will come from one pipe or chamber and the output will be directed toward another (Figure 8.4). Normally, both strokes are actively driven by a motor through a crankshaft and piston rod. While it's a fine arrangement, it's not the way a heart works. Hearts are made of muscle, and muscle is basically a tension-producing, shortening element. One half of each stroke, in practice the expansion half, is not directly driven by heart muscle contraction.

If a muscle is not disposed to expand actively, what other device can be provided so that the heart might beat more than once per lifetime? The muscles of our upper arms, for instance, work in antagonistic sets in which the active contraction of, say, the biceps not only flexes the forearm but extends the opposite triceps. Other arrangements abound. The halves of a scallop shell squeeze together as the result of the contraction of the large and tasty adductor muscle; they're reopened by an elastic pad just inside the hinge. Squid tentacles, elephant trunks, and many tongues have muscles intertwined and organized so that extension

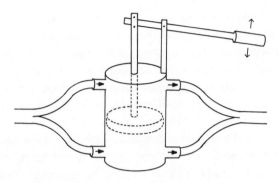

Figure 8.4. A double-acting push–pull piston pump; this one is arranged to be operated by hand. Note that there's no recovery stroke—pumping happens whether the piston is moving up or down.

can be accomplished by contracting radially and helically running muscles.[6] As mentioned earlier, spiders use a hydraulic system to antagonize their leg flexors. The outer mantle of a squid, the part that puts the squeeze on the water that squirts out to propel it, has a thin layer of extremely short radial muscle fibers. These act as mantle thinners and thus reexpand the mantle and reinflate the mantle cavity. Mantle also has elastic tissue, stretched by contraction of the squirting muscles, whose recoil helps reinflation.

Of course a heart might do nothing in particular and yet manage to refill. There's still some pressure available at the end of the venous system to push blood into the right atrium. A maximum gauge pressure of 15 mm Hg may be a little better than it sounds—because of the way mammals breath, overall pressure in the chest can be a bit subatmospheric. A more or less tacit presumption has long been that the phrase "nothing in particular" applies to heart refilling; indeed, strict application of Starling's law presumes no intrinsic expansion mechanism.

But, if not fame and fortune, then at least satisfaction and recognition come to the scientist who doesn't leave well enough alone, the busybody with no decent respect for illustrious forebears. To put the matter crudely, it now looks very much as if healthy hearts can suck on the veins that feed them. They don't do it with muscle fibers running in odd directions, as in the tentacles or mantle of a squid. Rather it comes back to all that stuff that makes heart a pretty tough meat—

fibers of connective tissue, both extensible elastin and nearly inextensible collagen. Every muscle fiber is quite literally laced up with fibers, so we shouldn't be surprised that heart has a lot of the character of a kitchen baster that rebounds into its spherical shape after one gives it a squeeze. In short, the natural tendency of either ventricle is to expand, thus pulling blood in from the atria. An excised mammalian heart continues to refill as well as to empty when kept in a container of appropriate solution. Even isolated pieces of muscle will spontaneously reextend, quite an unusual behavior for a muscle. Still, no one is getting a free lunch. That elastic material, whatever and wherever, must be deformed again, with attendent energy expenditure, when the heart muscle contracts in its systolic phase.

The same people who raised consciousness about elastic recoil, Thomas Robinson and his collaborators, made still another suggestion. They pointed out that in normal activity, hearts do a lot of moving around within the pericardial cavity. A contraction of the ventricles pushes blood up the aorta and pulmonary trunk. Blood has been accelerated, blood has mass, and the consequence is not at all different from the recoiling force you feel if you fire a rifle or hold a garden hose with water coming out the nozzle. Newton's third law, about equal and opposite forces, applies, as does the principle of conservation of momentum. Forcing blood upward pushes the heart downward, stretching the upper chambers and vessels lengthwise. That ought to reduce the pressure in the atria, at this point isolated by closed valves from the ventricles, and thus draw in blood from the vena cavas. They also suggest that in the diastolic upward rebound of the heart, the ventricles ought literally to swallow up blood from the atria. Again, it's not something for nothing, but it would certainly help get the job done.

Yet another mechanism, one that I found was probably used by squid to assist both elastic and muscular reexpansion of the mantle, is available. If you blow air or water across an upwardly convex surface, the pressure above that surface decreases. That drop in pressure sucks the surface upward; it's what we commonly call *lift*. It permits aircraft to fly and underlies all sorts of technology, both human and natural. The explanation invokes something called *Bernoulli's principle*, a bit of fluid mechanics normally of little consequence in circulatory systems. The faster a squid jets along, the lower is the pressure on the convex outside of its mantle, the more strongly the mantle is expanded, the faster and more fully the cavity beneath is filled with water, and the better poised is the squid to give its next jet pulse.

While the mechanism is unlikely to be of much consequence to any of the three hearts of a squid, it might just do good things for fishes. The main thing a fish does that takes a lot of power is fast swimming. Fast swimming should reduce the pressure near the widest part of the animal, and fish hearts are just inboard of this widest part. Thus the faster a fish swims, the more the heart ought to expand, drawing blood from both fore and aft in the body where pressures are higher. Still no free lunch—the lowered external pressure is a tax added to what systolic contraction must accomplish—but it would assist refilling when the need is greatest. In this regard, I was delighted when I first heard that fish (unlike us) increase stroke volume more than heart beat rate when they work harder, since this mechanism is one that works on stroke volume. I must admit that no one appears to have yet done the measurements necessary to find out if any fish really do take advantage of Bernoulli-assisted refilling when swimming rapidly. Also, I have to be candid and tell you that it won't do much for an animal living in air. No matter how fast the creature might run or fly, the effect will simply be too small to matter.

We clearly haven't come to the end of this expanded view of how a heart can work. We've seen that our hearts are attached to us mainly by vessels and are fairly free to move in the pericardial cavity. Maybe that movement is put to further use. When really pressed, a heart is usually inside a vehicle whose motion is anything but smooth. Recently, the functional morphologist, Dennis Bramble, has shown substantial coupling between locomotor and respiratory movements, especially in fast quadrupeds. Guts slinging back and forth act as pistons that help push air into and out of lungs. It's reasonably likely that the motion does something for hearts as well.

The movements associated with breathing clearly influence the performance of the heart. During a normal inhalation, you reduce the pressure in your chest by lowering the diaphragm and expanding the chest wall. Thus you draw air into the lungs. The reduced pressure draws blood into the various thoracic veins and into the heart, so the heart does a bit better than otherwise. During exhalation the opposite happens. The diaphragm is raised and the chest wall moves inward both increasing the thoracic pressure. As a result less blood is drawn into the thoracic veins and heart, and the heart works a little less effectively.

The interactions can be quite a lot more drastic. In what's called a *Valsalva maneuver* (described by Antonio Valsalva in 1707), a person tries hard to exhale with the glottis closed, that is, without letting the air get out. The result is a substantial increase in pressure in the thorax. This then greatly reduces venous return to the right heart, at least until a sufficiently high venous pressure can be built up to overcome the high-intrathoracic pressure. Thus little blood can go out to the lungs and little can return from the lungs, and the left side of the heart can't do much either. So don't do it—a Valsalva maneuver is unkind and even dangerous to your valuable body. Don't lift things so heavy that you have to hold your breath[7] to do so.

I might add a precautionary autobiological item. I was happily ignorant of the matter until I accidentally ran a test one nice fall afternoon. I had been hard at work hand-sawing firewood, and just after cutting it I lifted the last piece to put it in the pickup truck. The load was high and the tailgate up, so I probably held this large, unsplit chunk for more than the usual time. Anyhow, things weren't the same afterward, and I had 10 days in the hospital to contemplate both life and the alternative. A few weeks later, Knut Schmidt-Nielsen informed me about Valsalva and helped figure out what I'd done. Apparently worse even than the normal maneuver was doing it while breathless from exertion (which I certainly was). With almost no oxygenated blood coming back from the lungs, the coronary arteries were getting a pretty poor grade of stuff, and unusually little even of that. So the heart, that consumately aerobic muscle, was starved for oxygen just when the demands on it were especially high. If you're prone to heart trouble, the stunt is quite likely to bring it on. I'm a lot more careful now about how I treat my heart. I've largely given up hand-sawing logs, in part because the fun is somehow no longer in it. I'd guess that the prevalence of heart attacks among middle-aged shovelers of snow reflects similar cardiac abuse.

* * *

There's a moral here, perhaps a good one with which to end this particular chapter, with all the interactions that have appeared. It's a message that came up in Chapter 3, when we couldn't make a tidy distinction between nervous and endocrine systems. Organisms ought to work as integrated entities—it is at the level of the organism that the hand of natural selection should be most directly felt. Drawing tidy partitions between the various functional systems of organisms is all too handy, but it's likely to mislead. And we're likely to forget that a lot of distinctions are created simply to separate the chapters of a book or the courses of a college. Why should nature object to using a liver as a lung-assisting piston or to using a leg muscle as a circulatory pump? The seduction of reduction came up in the last chapter; these interactions show the advantage of keeping it at arm's length.

Notes

1. Lower figures for both bivalve mollusks and sponges mentioned are commonly cited. The figures here are the best numbers available since they were obtained with a totally noninvasive technique on undisturbed animals. In the case of the sponges, the animals were totally untouched and remained in situ on their original rocks in the ocean.

2. According to the *History of Hydraulics*, by Rouse and Ince, a very fine and readable account.

3. One sometimes runs into the term *minute volume*, which is how much a half-heart pumps in a minute. I don't much like the term, first because it mixes a unit (minute) with a dimension (volume), second because it really refers to a rate but makes it sound like a volume, and third because "minute" is an ambiguous word without the accent indicated.

4. Such a shunt does seem to exist in mammals, although any function in equalizing flow rates is uncertain. It's made up of the bronchial arteries, which convey oxygenated blood from aorta to the tracheal pipes of the lungs, and the bronchial veins, which convey most of the resulting deoxygenated blood into the *left*, not the *right*, atrium. (Lungs may oxygenate blood, but they do so in the alveoli, not in the tracheal pipes. Thus the latter need a proper arterial supply.)

5. It may perhaps be of no present relevance, but I particularly enjoyed Tom McMahon's most recent book, *Loving Little Egypt*, and not just as the novelty of an engineer's novel.

6. The best part of a nice article about this by Smith and Kier is that it

includes a picture of our former cat, Fred. For a can of tuna Fred was obliging enough to stick his tongue out repeatedly in spite of camera, flashgun, and my firm grip on his tail.

7. Breathholding while lifting seems natural to us. It has the advantage of helping to support the load hydro- and aerostatically, as in an inflated tire, taking some load off the spinal column.

9 *Moving Oxygen*

It's now time to redeem one of the promissory notes left fluttering in the first chapter. Almost everything since that point has been about mechanics in a narrowly newtonian sense, solid and fluid mechanics with the distinct odor of engineering. Such a narrow view simply won't suffice forever. As its most demanding task, our circulatory system distributes oxygen to our various cells to enable the cells to burn the fuel to release the energy to do the things we have to do. So our circulatory system is primarily a scheme for oxygen transport, and the specific mechanisms by which oxygen is carried are thus central to its story. The mechanisms may be chemical instead of mechanical, but mechanisms they remain. Oxygen, to state the obvious, has to be shifted from being a component of a gas mixture, air, to being dissolved in a liquid system mainly made of water in the lungs. It has to be transferred to another aqueous liquid system, blood. It must then be moved around by the now familiar circulatory machinery, and it has to be transferred from the blood first to extracellular and then to intracellular liquids.

Except for its carriage by the circulatory system, each of these transfers of oxygen involves the process of diffusion, described in terms of all those wandering inebriates back in the sixth chapter. It's crucial to bear in mind that net diffusive movement always takes material from regions of higher concentration to ones of lower concentration. What gives any semblance of directionality to the process is that concentra-

tion difference or a slightly more general equivalent of it that we'll euphemistically call "oxygen availability."

To view the matter with yet another analogy, let's imagine a rather
weird scenario. You're in one of two identical rooms interconnected by
a window. In your room a fly buzzes around distractingly. If you open
the window, then sooner or later the fly will fly into the other room,
something clearly desirable. But the other room contains nine flies, also
flying around. Now the chance of one fly making it through the window
is just the same as that chance for any other fly, since too few of them
are present to worry about such secondary problems as collision or col-
lusion. You might reason, then, that if you open the window you have
an even chance of improving your habitat. Your fly might depart or
another one might enter. If the former, swell; if the latter—well, you're
already distracted, so two isn't that much worse. So you open the window, and in a short time an unpleasantly larger number of flies have
moved over—apparently your reasoning was flawed. Where's the problem? The fly in the ointment is that each fly has the same chance of
going through the window, and more flies are present on the other side.
Therefore without anyone directing the operation, a *net* migration of
flies toward you will occur. And this *net* migration will happen until
the concentration of flies is, at least on average, the same in both rooms—
five per room. Net movement is obviously possible even with an undi-
rected diffusive scheme.

The properly perceptive or skeptical person will see a problem here.
Eventually the process stops. Flies keep buzzing through the window,
but net movement ceases. To make the analogy properly analogous to
oxygen transport, we have to replace one room with the great outdoors,
liberally endowed with so many flies that their concentration is unaffected by anything that might happen around our window—that's the
atmosphere of oxygen. We also have to give you, the inhabitant of the
remaining room, license to kill—that's an active tissue, using up oxygen. Now a statistically steady inward movement of flies continues in
perpetuity, as long as you steadily kill them. If you stop, the net inward
movement soon stops—as soon as the fly concentration in the room
builds up enough so as many flies are leaving as are entering. If you
kill them faster, the net movement increases since fewer ever escape
your room—at least up to the point where a fly has a very short expec-

tancy of occupancy. That's just the way oxygen moves from air to lungs to tissues as long as your cells consume the stuff, at least up to the limits of the supply processes, including diffusion.

The Need for Special Machinery

We might begin by asking how much oxygen a person needs. At rest, as mentioned in Chapter 1, each of us uses energy at a rate of about 80 watts. That takes a fairly specific amount of fuel, with only minor differences that depend on the mix of carbohydrate, fat, or protein you're consuming at any particular moment. Similarly, burning the fuel takes a fairly specific amount of oxygen, and the figure of 80 watts requires the use in every minute of about 10 fluid ounces of pure oxygen. Of course, you have to breathe a lot more than 10 ounces per minute. For one thing, the atmosphere is mostly nitrogen, with oxygen making up only about 21 percent of the molecules. So multiply by about five. Then you don't fully deplete air of its oxygen when you breath; you couldn't possibly do so. To extract all the oxygen, the concentration somewhere in the lungs would have to drop to zero. That's not possible as mammalian lungs are arranged since nothing can be diffusively transferred from a source that hasn't got any to begin with. An ordinary figure for extraction might be a quarter of what's initially present, so multiply by four. The 10 ounces is now up to about 1.5 gallons per minute. That's then the rate at which you process air when at rest. While it's about the same as the rate at which you pump blood, I think the coincidence of the rates is no more than that—coincidence.

Let's turn again to that rate of pumping blood, asking what you'll immediately recognize as still another of our what-if-it-were-otherwise questions. How much blood flow would it take to deliver that 10 ounces per minute of oxygen if we assume a blood with the properties of pure water? We can look up a figure for solubility of oxygen in water—at 15° C (59° F—a good average environmental temperature) about half an ounce (volume, not weight) of oxygen dissolves in a quart of water. Oxygen, as it happens, isn't especially soluble in water. It may be twice as soluble as nitrogen, but it's only a thirtieth as soluble as carbon dioxide. From that solubility, we might figure the amount of blood flow as about 2 gallons per minute, not far from what we actually pump. However, two terrible mistakes lurk here. First, the atmosphere, again, isn't pure oxygen, but is only a fifth oxygen, so there's less oxygen around to dissolve. That two gallons rises to 10 gallons a minute. Second, gases get progressively *less* soluble in liquids as the temperature goes up. That's why warm beer releases more gaseous carbon dioxide when exposed to atmospheric pressure than does cold beer. We're pretty warm creatures, with a body temperature of around 37° C or 99° F, so we're especially badly off, with still less oxygen dissolving in water at our body temperature. The 10 gallons rises to about 15. Thus were we to transport our oxygen merely dissolved in water, we'd have to pump that water at minimally[1] ten times the rate that we actually pump blood. I mentioned at the very start of the book that circulation costs less than 10 percent of our total energy consumption. Under these hypothetical circumstances, while it might cost less than our entire energy consumption, it wouldn't be very much less!

In the light of the last chapter comparison might be made with suspension feeding clams and sponges. The pumping rate calculated here, huge as it is, is still less than one body mass per minute. A sponge can do twelve and a scallop or mussel perhaps fifteen or twenty body masses per minute, but one probably shouldn't read too much into the comparison since the suspension feeders push fluid through shorter and broader channels with much less consequent viscous friction and pumping cost. Also, most of these suspension feeders live quiet lives and don't charge around spending energy on much else.

Hence Hemoglobin

When push has come to shove, nature has almost inevitably contrived ways to move oxygen in blood at higher concentrations than would be possible using simple aqueous solutions. Bloods come in quite a diversity of chemistries and colors, but almost without exception each contains some kind of complex organic molecule that has the interesting habit of grabbing a great deal of oxygen but not grabbing it very tightly. These are called *respiratory pigments* since all the common examples are colored substances. Sometimes the pigments are freely dissolved in blood, and sometimes they're contained in specialized cells. Sometimes the pigment is green and sometimes blue, but most often it's red. Our own respiratory pigment is hemoglobin, the one most widely distributed among animals (and present in a few plants); so, in a sense, it's the most ordinary of them all. We vertebrates, whether fishes, frogs, birds, or mammals, put our hemoglobin in specialized little bags, the red blood cells. The diverse invertebrates that use hemoglobin more often carry it loose in their blood plasma.

A few numbers ought to emphasize just what hemoglobin accomplishes. At body temperature, blood carries an amount of oxygen equal to about a quarter of 1 percent of its volume in ordinary solution, carrying it the way salt might be dissolved in water. So a liter of blood holds 2.5 milliliters of oxygen—a quart of blood with less than a tenth of an ounce of oxygen—in simple solution. That's the best it can do, taking up oxygen from the alveoli of the lungs at the concentration of oxygen present there. (The alveoli, because of the in-and-out character of breathing, contain air of roughly 13 percent, not 21 percent oxygen.) But enter hemoglobin. A liter of blood contains about 150 grams of hemoglobin, which combines with 210 milliliters of oxygen. So blood with hemoglobin carries over *eighty* (210 divided by 2.5) times as much oxygen as blood without it. In short, the amount of oxygen carried in simple solution is trivial compared to that carried in combination with hemoglobin. That, by the way, is the reason why an inadequate concentration of hemoglobin in the blood, what's called *anemia*, has such a drastic effect on one's capacity for sustained effort.

All of this suggests that for an active life any animal of decent size simply must employ some oxygen carrier. I loathe the expression "exception that proves the rule" unless one uses "to prove" in the less common sense of meaning merely "to test." As it happens, there is a very instructive exception that tests the limits of necessity for an oxygen carrier. A few antarctic fish entirely lack hemoglobin or any other respiratory pigment. They're known as "ice fish," not because of biogeography, but rather because of appearance. These are very nearly transparent creatures, the only known transparent adult vertebrates. Transparency is certainly the ultimate in cryptic (hiding) coloration, so one doesn't have to look far for some adaptive advantage. The "test" happens when we examine their habit and habitat. These are sluggish animals, goners if they're spotted by any proper predator, and thus they should not need a particularly copious oxygen supply. (One must by their existence guess that the local predators are predominantly visual hunters.) Furthermore they live in the coldest of waters, never over 37° F (3° C), at temperatures at which nearly twice as much oxygen dissolves in water as at our body temperature of 99° F (37° C). Finally, the local water is well oxygenated, so they can load up their blood nearly to full atmospheric levels. Thus the playing field is well tilted in their direction. Still, it does seem a bit strange that transparency entails such drastic sacrifice, and that nature hasn't contrived a colorless oxygen carrier. The main point, of course, is that life's pretty restricted without an oxygen carrier.

At this point I have to introduce a slightly specialized device in order to talk about all the neat functional features of hemoglobin. The stuff takes up oxygen in one place and drops it off in another. What matters are, first, how much oxygen it can carry, and, second, what the conditions are for taking it up and dropping it off. The first factor is what we've just been talking about—*oxygen carrying capacity*. The second, the business of loading and unloading, what we might call *oxygen avidity*, is easiest to deal with if kept separate from the first. One of the persistent prejudices one acquires in doing science is a strong preference for dealing with variables one at a time. The most common procedure in an experiment is to keep everything constant except one factor so as not to get confused or misled. (Sometimes that procedure itself proves either misleading or impractical, so I hasten to add that I imply no absolute rule.) The device for presenting information on loading and unloading is a graphic scheme called a *dissociation curve*, in which the *relative* amount of hemoglobin that is carrying oxygen is plotted against the concentration of oxygen to which the blood is exposed.

Figure 9.1 gives such a graph, with a set of axes that aren't exactly self-explanatory. Going across is the concentration of oxygen present. Now the amount present, in the usual sense of mass or volume, depends on such things as whether oxygen is free in air or is dissolved in water or is carried by some chemical. Thus it would get terribly confusing to put any of our familiar measures of amount on that horizontal axis. We'll use a scale we'll call "oxygen availability," evading a lot of complication available, if need be, from any biology text.

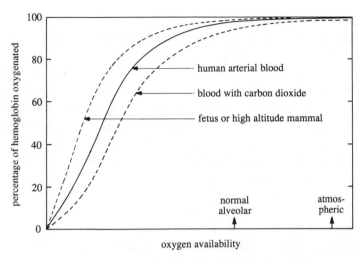

Figure 9.1. Dissociation curves for three sorts of hemoglobin. The curved shape of the lines and the left or right shifts due to change of organism or circumstance are what matter most here.

Going upward is the relative amount of oxygen bound to the respiratory pigment. This vertical axis runs from none to the amount bound if every hemoglobin molecule had all it could hold; we'll call the scale "percentage of hemoglobin oxygenated." What are most important in the operation of one of these carriers are the conditions under which it will take up ("associate" with) oxygen and under which it will let it go ("dissociate") again—that's what a dissociation curve shows. To repeat, we deal with other factors—how much hemoglobin is in the blood and how much oxygen each unit of hemoglobin can maximally carry—separately.

The first thing that strikes one about a dissociation curve is that it's just that—curved. By contrast, a simple solution of oxygen in water gives an ascending straight line beginning at zero availability and zero loading. The fact that the curve for blood starts by rising with increasing steepness as one moves to higher availabilities means that having a little more oxygen around results in a lot more being carried by the blood. Or, looking the other way at this left portion of the curve, as the local oxygen availability drops (moving leftward), the blood will release a disproportionate fraction of what it's carrying. That's very nice! Work a tissue, a muscle perhaps, a little harder; run the oxygen level in the tissue down a bit; and the blood will release a lot more oxygen for use by that tissue.

At the other end the curve levels off toward a horizontal line. While it might take a very high concentration of oxygen to load the blood fully, very nearly full loading happens at quite a lot more ordinary levels. Thus the concentration of oxygen in the air in our lungs is enough to bring blood to 97 percent of the maximum amount of oxygen it can carry. You can get full loading if you can get the pulmonary concentration up to about that of atmospheric air, but the 3-percent gain isn't worth a lot of huffing and puffing in anticipation of some heavy activity.

The mountainlike shape (a *sigmoid*) is a fairly general feature of dissociation curves, even those in which the respiratory pigment is something other than hemoglobin. The overall steepness of the curves and their positions in the left-right direction are the variables. The left-right position turns out to be of considerable interest and provides lots of good tales to relate. As a general matter, a shift of a dissociation curve to the *left* implies that the particular kind of respiratory pigment will load closer to its ultimate capacity with less oxygen around, that the blood has a higher avidity for oxygen. At the same time, tissue oxygen levels will have to get lower for the blood to release as much. A shift to the *right* has the opposite consequence—the blood has a lower avidity, so a higher level of oxygen is needed for loading, but the blood has a greater willingness to unload. The curve given for human blood is quite an ordinary mammalian curve both in shape and position; it makes a nice base for the comparisons that follow.

Fine-Tuning the Oxygen Carriage

In a very real sense the arterial dissociation curve given in Figure 9.1 is a laboratory artifact. It might be taken as a good example of how one can be mislead by our nicely "scientific" practice of permitting only one factor to vary at a time. From the curve it looks as if even at an oxygen availability a fifth of that in the lungs, hemoglobin will still be about 70 percent loaded. Anything approaching complete unloading would require very low local oxygen levels. In reality something else changes, and unloading and loading don't happen under equivalent conditions.

The one consistent result of the utilization of oxygen is the production of carbon dioxide. It's nearly a perfect one-for-one molecular trade. Thus the production of carbon dioxide is the best signal that oxygen is in demand, and a persistent increase in production of carbon dioxide requires an increase in delivery of oxygen. It happens that carbon dioxide interacts with hemoglobin molecules in a most appropriate manner. *The higher the carbon dioxide concentration the further the dissociation curve is shifted to the right.* In other words, an increase in carbon dioxide quite without any change in local oxygen availability makes hemoglobin release additional oxygen. At that oxygen availability of a fifth that of lung air, the presence of a lot of carbon dioxide can drop the loading level of hemoglobin from the 70 percent just mentioned to as low as 20 percent. That gives an additional unloading of fully half the oxygen passing in the blood through the tissues. The phenomenon is known as the *Bohr shift*, after Christian Bohr,[2] who discovered it early in this century. An altogether admirable arrangement!

Another tidy trick. Hemoglobin isn't a single species of molecule, but is rather a whole set of variations on a theme. In chemical terms, something called a *heme group* is the immediate business end; but its precise dissociation curve depends on a variably constituted associated protein called, unimaginatively, *globin*. (Actually a full molecule of the basic vertebrate hemoglobin is a tetramer—four of each.) An animal normally makes a single version of hemoglobin, with its characteristic loading and unloading behavior. That's trouble, if of a subtle sort. A mammalian fetus doesn't breath air, but instead obtains oxygen through the placenta, loading up with the stuff from the maternal blood. So fetal blood is yet another step further along the concentration gradient that drives diffusive exchange. As fetuses we made a different hemoglobin, one with its dissociation curve shifted to the left. Thus we could load up to the same level at a lower oxygen availability, an availability characteristic of maternal arteries instead of air in lungs. After birth we gradually stopped making the fetal kind of hemoglobin and slowly replaced it with the grown-up sort. One is again and again astonished at the ramifications of birth!

And yet another trick. Everything so far has presumed that an animal lives in a sea level atmosphere. Lots of animals never know such rich air, living their lives at pressures substantially lower. (Atmospheric pressure drops to half its sea level value at about 5500 meters or 18,000 feet.) Well-adapted high-altitude mammals such as llamas do the sensible thing and make hemoglobins more attuned to life up there. Their hemoglobins have dissociation curves shifted to the left of ours, roughly in the position of human fetal hemoglobin. Even members of various subspecies of a kind of mouse trapped at different altitudes have been found to have hemoglobins with different dissociation curves. Birds also do the same thing, with those species that cross high mountain ranges on migratory flights having leftward-shifted dissociation curves. Still, neither birds nor mammals set any records for left-shifted curves. A far more extreme case is that of an especially large and ugly lugworm, *Arenicola* (Figure 9.2), that lives in a burrow in the muddy sand of coastal mud flats. It gets oxygen by pumping water through the burrow when water is available above it, but when the tide is out on a warm afternoon, things can get pretty stagnant and anoxic. Thus it spends a fair fraction of its life in a habitat in which oxygen is scarcer than on any mountain top.

Unfortunately, the adaptation is basically genetic. It seems that no amount of exposure to high altitude can persuade human cells to make a hemoglobin with an appreciably greater interest in loading oxygen. The dissociation curve of hemoglobin can be altered by changes in the concentration of substances normally present in our cells (as we saw for carbon dioxide), but we don't play that game to any great extent either. Nor, as far as we know, does any other species. All of which is most peculiar in that every one of our cells retains the information needed to make fetal hemoglobin, which would be nice to have for life at high altitude (for all except pregnant women). Sorry, mountain climbers.

That isn't to say, I hasten to add, that humans make no physiological adjustments to life at high altitudes. People who grow up at high altitudes develop larger lungs, so more area is available for diffusion from air to blood. Substantially supernormal numbers of red blood cells occur both in high altitude natives and in acclimatized lowland people who've moved up in the world. Proliferation of capillaries, thus reducing the diffusion distance from blood to cells, is yet another element in

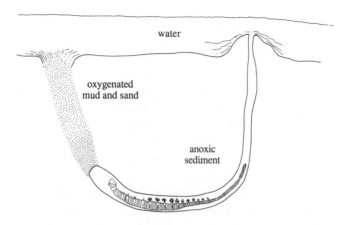

Figure 9.2. The lugworm, *Arenicola*, in its burrow. The animal itself may be as much as a foot long—but it's only repulsive and not harmful.

acclimatization. If there's any change in the oxygen affinity of hemoglobin, it's a decrease—a slight shift to the right of the dissociation curve—facilitating release of oxygen in active tissues.

The subject of body size loomed large back in Chapter 6, when the consequences of the relatively higher metabolic rates of small creatures were briefly explored. At the cost of a little violence to evolutionary history by using us big creatures as a norm, we can recognize several compensations for small size. In particular, capillaries are closer together, and blood doesn't hold oxygen quite as tightly. The latter, again, means that the dissociation curve is shifted to the right. Both factors, greater capillary density and looser oxygen binding increase the availability of oxygen to the more active cells of small animals. Remember, it's the steepness of the concentration gradient that matters for diffusion—a difference in concentration divided by the distance over which that difference occurs. Small animals have both a larger numerator and a smaller denominator of that fraction than do large ones; thus the gradient is much steeper and diffusion concomitantly facilitated.

In moving up to a higher altitude, a person does what amounts to a change toward the situation normal for a smaller animal. While a small animal may have a normal oxygen supply, it has a greater oxygen demand. Thus (to change viewpoint), it faces a situation analogous to that of a larger animal at a high altitude, where demand may be normal, but supply has dropped.

Small animals have yet another variation on the theme—their dissociation curves show a greater Bohr shift—a little carbon dioxide causes a lot of extra oxygen release by their hemoglobin. In fact, there's no threshold of smallness, but rather an inverse relationship between the magnitude of the Bohr shift and body size. Thus not only are the curves of small animals shifted rightward, but the Bohr shift effectively pushes the curves rightward still further, greatly increasing the oxygen supply as tissues get metabolically fired up.

In all this talk about the diversity of dissociation curves one shouldn't lose sight of a crucial characteristic of the overall process. Most of the time we're interested in nice, stable chemicals. If you burn a substance such as sugar or starch, the combination with oxygen that we call either burning or aerobic metabolism releases energy and leaves stable products—specifically water and carbon dioxide. If you combine hemoglobin with oxygen, as happens in lungs or gills, the process neither absorbs nor releases much energy, and the product, oxygenated hemoglobin, isn't at all stable. Only a minor shift of circumstances, a lower oxygen concentration and perhaps a little extra carbon dioxide, suffices to reverse the process and break up the product. The weakness of the combination is functionally critical.

It's possible to link hemoglobin much more strongly to something, and the possibility is best avoided like, quite literally, the plague. It's a grave matter that can do you in faster than bubonic plague during the worst days of the black death. The most common and dangerous case is combination with carbon monoxide. Carbon monoxide happens to be odorless, tasteless, colorless, and a common byproduct of incomplete combustion. Its combination with hemoglobin is distressingly strong, so loading your hemoglobin with carbon monoxide is a little like filling dump trucks with solidifying cement. To make matters worse, the effect is somewhat cumulative, and the combination makes blood even redder than usual, so victims sometimes appear in what might be called the pink of health. Treatment for carbon monoxide poisoning is a bit trickier than administering an emetic or antidote. Breathing pure oxygen takes maximum advantage of simple aqueous solution, and addition of a little carbon dioxide stimulates more rapid breathing. Fancier therapies include supplying oxygen at superatmospheric pressures with the victim in a hyperbaric chamber.

And What About Carbon Dioxide?

All the talk so far has been about the problem of getting oxygen from gills or lungs out to the rest of the cells of a body. All that oxygen then ends up burning carbon-containing compounds and producing water and carbon dioxide. The water we can afford to ignore (although it may be an appreciable part of the water supply of some extremely drought-adapted animals.) The carbon dioxide, however, must get back out to the atmosphere. Any local accumulation of carbon dioxide is most unpleasant, since it combines with water to make carbonic acid,[3] which makes the blood more acid; and an excessively acidic blood is vastly worse than any overly acidic stomach. So what should we say about the problem here? First, the chemical interactions between carbon dioxide and the other constituents of the blood are a complex lot. That in itself is a good reason to sweep the matter under the table in a book that's trying to stay as clear of chemistry as possible. Second, carrying carbon dioxide isn't as much of a problem as carrying oxygen anyway, since it's thirty times as soluble in water as is oxygen. (The real problem is controlling the acidity that the solution represents.) In fact, a lot of the carbon dioxide is carried in combination with hemoglobin—part of the same interaction that produced the Bohr shift.

Alternative Pigments and Adaptive Tales

A recurrent theme here, sometimes tacit and sometimes explicit, is the elegant tuning of the design of animals to the demands placed on them. The results of evolution are impressive both in the diversity of forms we see and in how well all that diversity manages to do what has to be done. The tuning is so clearly effective that faced with some newly uncovered novelty we ordinarily presume that the arrangement must "make sense" in a functional context. Strictly speaking, that hypothesis isn't at all easy to defend as dogma, however good a guide it might be to further investigation. A developmental context—making sense in terms of how something comes about as an animal grows up—presents no theoretical problem since every adult feature must have developed in a lawful manner. An evolutionary context—making sense in terms of ancestral continuity and incremental modification—isn't much worse as a philosophical problem, even if we usually have little more than speculation to work with. It's the functional context that presents the real problem—no law demands that each feature have a function, much less that it function uniquely well. But the assumption of functional utility works so well and scratches our prejudices so seductively that it takes a real act of will to detach ourselves now and then.

You'll see why that cautionary preamble is necessary as we look at how several features of the oxygen-distributing equipment vary from organism to organism. First is the choice of respiratory pigment. Hemoglobin, if we include all its minor variants, is a very widely distributed kind of respiratory pigment. As we noted earlier, it appears in at least a few species in every one of the important animal categories (*phyla*); beyond that it's even present in the nitrogen-fixing root nodules of some plants. It seems to have been evolved as a respiratory pigment on a large number of occasions, but that shouldn't be surprising since this very complicated molecule didn't have to be cooked up from scratch each time. As it happens, quite similar molecules are involved in the internal chemistry of almost every kind of cell. What might raise an eyebrow or two among us properly red-blooded sorts is that while nature has come up with other respiratory pigments, none seems to do anything that hemoglobin can't do at least as well and usually better. Many worms use hemoglobin, but a few marine species use a green pigment; a few others, also marine, use a violet one. Stranger still, some of the green worms make some hemoglobin as well. A more common respiratory pigment is a light blue one; it occurs in a lot of different kinds of mollusks, including some snails and the cephalopods. In addition, it's the oxygen carrier for many crustaceans. For reasons that obviously transcend the scientific, lobsters have been the most popular animals from which to obtain samples for experimentation.

To set the stage for a broader comparison, consider what the different hemoglobin-containing bloods can do. As we saw, our blood can carry an amount of oxygen equal to 21 percent of its volume (210 milliliters in each liter was the datum given). That's a typical value for birds and mammals. The other vertebrates usually carry a little less— 5–15 percent would include most of the known cases. That giant earthworm mentioned back in Chapter 4 isn't much different, carrying about 14 percent oxygen in its blood. As is the wont of worms, its hemoglobin is dissolved in its blood instead of being carried in cells, but that seems to make no huge difference in carrying capacity. No blood using any other respiratory pigments exceeds even the lowest of these figures. If the measure of a respiratory pigment is the relative amount of oxygen that a blood using it can carry, then hemoglobin is better than anything else we know about. Also, as we've seen, it certainly can't be faulted for any lack of versatility.

Nonetheless, despite its ubiquity, hemoglobin is clearly not the universal choice of respiratory pigment. Our inclination, from long habit and past success, is to look for reasons why each of the other pigments might have an advantage under some specific circumstance. The inclination is especially appealing when neither a developmental nor an evolutionary context gives much of a hint about the odd distribution of the pigments among different animals. The reasons, however, have not emerged despite a lot of work and thought. The difficulty of hitting on a persuasive functional explanation, a "why" in a functional context, is perhaps best illustrated by turning to a complex system that doesn't use hemoglobin and about which we know quite a lot.

We turn, then, to our recurrent alter ego, the octopus, and to its fellow cephalopods. They rely on the blue, copper-based hemocyanin instead of the red, iron-based hemoglobin. The blood volume of a squid or octopus isn't much different from ours—about 5 percent of body volume—and cephalopod blood is about as loaded with hemocyanin as ours is with hemoglobin. But that blood carries an amount of oxygen of at most about 5 percent, not 15 or 20 percent of its volume. The system works well enough at rest. Relative to its weight, a 1-kilogram octopus consumes oxygen at about a third the rate of a 70-kilogram human. Similarly, its cardiac output is about a third of our 1.5 gallons per minute, again relative to its weight. A squid, a much more active animal, has a cardiac output as much as three times ours and thus nine or ten times that of an octopus. (I must mention that squid prove to be exceedingly difficult laboratory animals, so we know more about octo-

puses.) These creatures do appear to have highly competent sets of pumps
and pipes.

Or are they really so good? Perhaps using humans for comparison
exaggerates their capabilities. As noted in both Chapter 6 and earlier in
this chapter, small animals use oxygen at higher rates relative to their
weights than do large ones. The resting oxygen consumption of a 1-
kilogram octopus is about ten times lower than that of a 1-kilogram
mammal. The oxygen consumption of a squid is about the same as that
of a mammal of the same size. Still, these aren't particularly bad numbers.

They get into real trouble, though, when exercising. By our standards, both octopus and squid have tired blood; neither can sustain
high levels of activity under either initiative or duress. Perhaps the most
relevant variable for a comparison is something called *metabolic scope*
by comparative physiologists and *mets* by exercise physiologists working
with humans. By either name, it's the factor by which resting oxygen
consumption can be multiplied in maximal sustained activity.

A typical mammal can raise its resting oxygen consumption tenfold
when active—we say it has a metabolic scope of 10, or we rate it at 10
mets. Along with dogs, horses and such, we're exceptionally good at
sustained exercise, perhaps reflecting the hunting techniques of distant
ancestors. Even I, a middle-aged veteran of heart trouble, carry a rating
of 17 mets on my identification badge at the rehabilitation clinic. Properly trained runners have ratings well up in the 20s. Birds may be as
good, but the figures available are around 15. (Of course one has to
realize that getting these data requires that an animal exercise in a fashion not very different from its normal activity so as to use its best-
developed muscles, that the activity has to be done while its oxygen
consumption is being monitored, and that the animal has to be willing
to do its best for some human with whom it may not enjoy special
rapport.[4])

Then there are the poor cephalopods. The metabolic scope of an
octopus is only a little over twofold. Even that number may give an
overly generous view—remember that the resting energy consumption
of an octopus is about ten times lower than that of a mammal of the
same size. No surprise about this last datum, perhaps, since an octopus
doesn't bother to keep body temperature above that of the surrounding

water, and it's neutrally buoyant to boot. The result, though, is that it moves a lot of blood relative to its rate of oxygen use at rest, and that it consequently can't do much better in activity. An octopus may make a sudden jetting movement, but it can't keep any decent pace.

A squid does rather better, with its resting rates of oxygen use of ten times that of an octopus. While that may sound impressive, the larger picture again isn't so glorious. The metabolic scope of a squid seems to be between 1.5 and 2.0—lower even than that of an octopus. Squid manage, it seems, by using a very fine pumping system and running it nearly full bore nearly all the time. A squid can get up to very high speeds, possibly hitting 20 miles per hour, but that's just a quick set of jet pulses when pursued by some pair of vertebrate jaws. Afterward it's left gasping, or whatever word might apply to a breathless cephalopod.

It's hard to avoid the deep suspicion that these animals would be very much better off with a good dose of hemoglobin, even of the sort used by that giant earthworm. We've made much of the sophistication of their hearts and arteries—are we engaging in a subtle sort of self-deception? Maybe we're just looking at a set of compensatory adjustments by creatures trying to run but stuck with a terribly pedestrian respiratory pigment. Perhaps those with bad blood need be good at heart!

In a sense, the cephalopods may be in a trap set by their ancestors. Octopuses, at least, are about as intelligent as anything outside the mammals. They're big-brained creatures that deal with complex habitats, that learn many tasks with ease, that have the manual (or, should we say, tentacular) dexterity to do complex tasks, that have the sensory equipment to go with that dexterity and versatility, and that even in our crude view have individual personalities. But octopus and squid are still mollusks, without either moltable exoskeletons like arthropods or growing bones like vertebrates, and they have (albeit much modified) a fundamentally molluscan circulation. Martin Wells, a long-time student of the cephalopods, puts the matter well in a review of the system.

> The basic design of the molluscan circulatory system was laid down long before large, active members of the group emerged. It was well suited to a slow-moving animal . . . but a disaster for large, high-metabolic rate invertebrates, competing with fish and other vertebrates. Natural selec-

tion works on the material available, and the cephalopod circulatory system is a miracle of making the best of a bad lot.

One more bloody tale to keep us suspicious of facile stories about the perfection of the design of organisms. Vertebrates, as mentioned, put their hemoglobin in cells—red blood cells, we call them. Fish, amphibians, and reptiles use cells of the conventional sort—reproductively competent little things with nuclei of the kind you might meet in some beginning biology course. We mammals get a bit fancier. Our red blood cells aren't, by the strictest definition, cells at all, and purists call them *red blood corpuscles*. When circulating around transporting oxygen with their hemoglobin, they lack nuclei and much of the other conventional cellular machinery. They have a limited lifespan (about 100 days in humans) and have to be continuously produced by the bone marrow of sternum, ribs, skull, and vertebrae.

The scheme certainly sounds like a good way to get the maximum amount of hemoglobin into the blood with the least number of red blood cells. It certainly makes sense as something crucial or at least highly desirable in converting a sluggish reptile into a hot and active mammal. We'd almost certainly give this most satisfying explanation—but for one thoroughly embarrassing circumstance. Creatures exist—we call them birds—that have oxygen delivery systems at least the equal to anything the mammals have come up with. A climber gasping for air atop a high mountain might well envy a bird flying easily over the peak. Nothing else involves sustained oxygen use at the rates characteristic of flapping flight. Yet birds have normally nucleated red blood cells. Clearly loss of nuclei is not the sine qua non of a high capacity for transporting oxygen.

Notes

1. The "minimally" recognizes that I've ignored such problems as maintaining a concentration difference to drive oxygen from air into blood and from blood into cells.

2. The father of the more famous Niels Bohr, known for the Bohr model of the atom, and the teacher of August Krogh, whose name has come up several times so far.

3. Which is why a flat soft drink tastes sweeter than one still carbonated. As the carbon dioxide bubbles out, carbonic acid forms more carbon dioxide, so the acid that earlier offset some of the perceptible sweetness gets depleted.

4. On one occasion, for instance, a cheetah decided that putting the bite on the investigator might be more amusing than running on a treadmill.

10 *Moving Heat*

Here's a problem. Imagine a solid body of which each little bit produces heat, with the heat produced at the same rate wherever the bit happens to be located. Heat production is uniform throughout the body—will its temperature likewise be uniform? As it turns out, under most circumstances the temperature will be far from uniform. The inside will be hotter and the surface will be cooler than the average temperature of the body. That the body produces the heat makes the problem only slightly less ordinary than what happens when you take some long-cooked casserole out of the oven to cool. It may be at a uniform temperature when it's taken out, but it soon gets cooler around its outside than at the core. If the soup is too hot, you sip first from the edge.

In slightly more formal terms, we can say that, while heat production happens throughout the volume of the body, heat loss to the environment happens across the surface. Heat thus has to be continuously transferred from center to periphery. This presents no great difficulty since losing heat at the surface will lower the temperature there. The middle will therefore be hotter, and heat normally moves down a temperature gradient (gradients again!); it goes from warmer to cooler. The process, *conduction*, works just the same way as does diffusion; even their mathematical descriptions are identical but for the replacement of a single factor, diffusion coefficient, with another, called *thermal conductivity*. The process of heating works the same as does cooling except that the gradient runs the other way, from hot periphery to cool core.

Put the roast in the oven, and it picks up heat across its surface, only very slowly approaching a uniform temperature (which you don't let it reach since it would then be horribly well done!)

We mammals and birds engage in just this kind of continuous heat production. For that matter, all organisms do it, but we simply produce enough heat to get body temperature (usually) well above the temperature of our surroundings. We maintain a high and fairly constant temperature—as we'll see shortly, the exact value depends on the particular kind of organism. Not only is the temperature fairly constant from moment to moment, but it's also reasonably constant throughout the body. Constancy, though, means the *absence* of just that temperature gradient that we argued was necessary to get heat from core to periphery, but which was automatically established under ordinary circumstances. Thus a uniform internal temperature is emphatically not the ordinary thing, and its existence needs some special explanation. If conduction ruled, constancy could only be achieved if all the heat we produce were to be generated by cells and tissues located at or very near our outer surfaces.

In fact, surface heat production is clearly not at all what goes on; reality is skewed in quite the opposite direction. Good data exist for heat production by the different parts of a human body. This data can be easily converted to heat production relative to organ weight—the fraction of the body's heat production divided by the fraction of the body that the organ makes up. Were heat production uniform each body part would give a value of 1. At rest, though, skin has a value of ¼—it's metabolically inactive stuff. Muscle is only a little higher, with a value of ⅓ (at rest, of course). Lungs, brain, and the digestive organs are far more active, with relative heat productions of 5 to 9. In short, each produces five to nine times as much heat as would be expected on the basis of its weight. Even they prove relatively quiescent though. Kidneys, busily keeping blood composition regulated, produce seventeen times the expected heat. The winner, far exceeding even that renal rate, is our present chief protagonist, the heart, generating fully twenty-four times its share of body heat. Perhaps Galen (Chapter 3) wasn't entirely unreasonable in ascribing production of the body's heat to the heart!

Overall, the big heat producers are centrally rather than peripherally or even uniformly distributed. Thus, the constancy of temperature from place to place in the body is even more special than implied earlier—something other than conduction of heat simply must be involved. What's going on is another mode of internal heat transfer—*convection*. Blood moves through hot organs, is itself heated, and then gets pumped to cooler climates such as skin and lungs, where it loses the heat. For animals in which body temperatures are substantially above ambient temperatures, convective heat transfer is an important function of circulatory systems.

Convection through Circulation

Both diffusion and conduction absolutely require gradients—in the first of concentration, and in the second of heat—so uniformity of neither concentration nor heat can be achieved unless an infinite amount of time has elapsed. In the near-identity of diffusion and conduction, however, lies a common way of escaping their limitations—the shift to forced fluid movement, to convection. Forced convection still takes a gradient (the physicists serve no free lunch), but it uses a pressure gradient, maintained by a pump. It thus circumvents any requirement for temperature or concentration gradient.

(To tell which of the processes, diffusion or heat conduction, is in worse need of convection would require that we compare their speeds, which is a little tricky. We can't easily equate concentration and temperature differences nor the relevant material properties, diffusion constants and thermal conductivities. Still, a loose statement isn't too risky a noose, so I'll just declare them both slow and thus seriously constraining, at least if organisms are intolerant of really high internal temperatures and persist in rejecting metals as structural materials.[1])

It's likely that the best feature of this scheme for convective heat transfer is one of those "automatic" adjustments such as that provided by the Bohr shift that figured in the last chapter. Recall that if a tissue produced extra carbon dioxide, it automatically got an increased ration of oxygen. Here, if an organ has a high rate of heat production it automatically gets a more copious blood supply. The automatic feature is simply a consequence of the fact that heat production is an inevitable result of using oxygen. When we say we burn some metabolic fuel with oxygen, the word *burn* isn't nearly as metaphorical as you might have guessed. Measuring the rate of heat production is a perfectly good way to follow energy consumption, more commonly called *metabolic rate*. As a matter of convenience we usually measure oxygen utilization, but heat production in fact has certain theoretical advantages.

A Little about Being Warm

This is a good point for a slightly more cosmic look at some aspects of thermal biology. Birds and mammals were earlier declared "warm blooded"; by implication all other creatures are cold blooded. That's really a bit glib, one of these cheap dichotomizations that we humans, even those of us who profess science, find nearly irresistible. The underlying generalization is that birds and mammals, without exception, use internal heat production (burning extra fuel) to maintain body temperatures that are fairly constant and most often above the temperatures of their surroundings. While all birds and mammals do it, however, they all don't always do it. A greatly decreased temperature is one of the hallmarks of hibernation, and tiny creatures such as shrews and hummingbirds maintain high body temperatures only during the daytime. Conversely, lots of other creatures take active steps to keep warm. The temperatures of the flight muscles of large and moderate-sized insects are maintained well above ambient levels during flight; for some insects a period of preliminary shivering or slow flapping is needed to get up to take-off temperature. Reptiles commonly lie exposed to sunlight in postures arranged to get themselves above air temperature, and some insects do the same. Even some large fish, as we'll get to later, can be at least in part "warm blooded," using specific devices to retain the heat produced by their muscles as they swim rapidly.

There must be some benefit to keeping warm since it almost always takes either time or energy. The phenomenon probably has two elements, warmth and constancy, which both confer advantages. These advantages are almost certainly related to a peculiarity of most chemical reactions; (other things being equal) they proceed faster at higher temperatures. And not just a little faster. Raise the temperature of a system by 10° C (18° F), and the chemistry usually speeds up between two-

and threefold. Hotter therefore means faster, and the faster predator catches the prey. How much faster, though, varies a bit among the hundreds of reactions that take place continuously within every cell. Avoiding any coordination problems from all that variation in the speeds of reactions may be the main benefit of keeping temperature relatively constant as well as high.

The cost of keeping a high and fairly constant body temperature, however, varies greatly from animal to animal. For one thing, the particular temperature chosen isn't at all strongly correlated with the temperature at which an animal lives. It isn't as though we have a bunch of creatures each keeping its temperature above that of its surroundings by the same amount. Nor does the temperature reflect some standard cost of keeping constantly warm—the cost is clearly greater for an animal that lives in a very cold climate than for, say, a mammal on the floor of a rain forest. Instead, the temperature chosen seems to reflect in part specific ancestral tradition and in part some kind of general constraint. Most mammals chose something between 36 and 38° C, (97 and 100° F) and most birds are between 39 and 41° C (102 and 106°F), but an aberrant and more reptile-like mammal, the egg-laying spiny anteater, regulates at 30–31° C (86–88° F). The general constraint that seems to set the temperature is something related to the stability of the proteins of the body. Enzymes, which are proteins, may work faster at higher temperatures, but they typically become unstable and inactive at disproportionately increasing rates as temperatures climb above about 40° C (104° F).[2] Somehow it seems as if nature has decided that if body temperature is to be kept constant, it might as well be about as high as possible, thereby nudging the spectre of protein instability whatever the cost.

Some of the other factors that determine that cost are more subtle than ambient temperature. One is the medium in which an animal lives. The cost is greater for a mammal or bird that swims in water than for one that lives in air—we make much more effective thermal contact with water than air. A naked and sedentary human is comfortable down to an air temperature of about 27° C (81° F) but gradually gets chilly in water below 35° C (95° F). Fur and feathers, fairly cheap insulation, are of limited value in water (although air-filled furs or feathers seem to work for brief immersions). A layer of fatty tissue—blubber—works well, but it takes a lot of energy to synthesize. An arctic seal, for instance, may have 50 percent of its mass invested in blubber, an energy investment greater than that of the rest of its body put together.

The most extreme factor in the cost account is body size. Heat, remember, is produced throughout the volume of an animal, but it's lost across the outer surface—precisely the same situation we saw with oxygen acquisition and utilization. The crucial and inescapable factor is a geometric one. A bigger animal has more inside relative to its outside than does a small one; a small one has more outside relative to its inside. Consider two cubes (Figure 10.1), one ten times larger on every edge than the other. The smaller has a total *surface* of 1 unit times 1 unit times 6 faces or 6 square units, the larger of 10 x 10 x 6 or 600 square units. To be tediously explicit, the larger has 100 times the surface of the smaller. At the same time, the smaller has a *volume* of 1 x 1 x 1 or 1 cubic unit, while the larger has a volume of 10 x 10 x 10 or 1000 cubic units. Again to beat the moribund horse, the larger has 1000 times the volume of the smaller. Here, finally, is the point. *The smaller has ten times as much surface relative to its volume as has the larger*— relatively more outside and less inside. The reasoning works equally well for any pair of geometrically similar objects—spheres, a set of cylinders twice as long as wide, and so forth.[3] What's implied is that when it comes to heat production, the larger, volume-rich animal, is better off. When it comes to heat exchange (loss, mainly), the smaller, surface-rich one is in a better position.

For a big animal, then, staying warm enough isn't that big a deal. For example, a bear doesn't really hibernate, but mainly sleeps through the winter. When you're that big, heat loss is bearably low without much reduction of body temperature. As it happens, a sleeping animal is more easily aroused than one in true hibernation. I've heard that an intrepid investigator became aware of this last fact when attempting to measure the core (= rectal) temperature of an overwintering bear in its cave. Perhaps dinosaurs were warm-blooded; as a physiologist, I'm not especially impressed since they ran toward largish bodies.

Conversely, heat loss is a problem for small mammals and birds even in relatively warm climates. As noted earlier, shrews and hummingbirds let their body temperatures drop every night. They have so much outside relative to their insides that the energy needed to keep warm through the night would presumably be unaffordable. So the

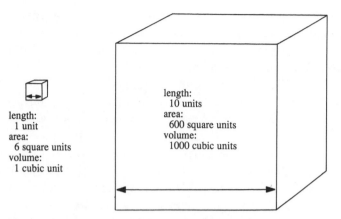

length:
 1 unit
area:
 6 square units
volume:
 1 cubic unit

length:
 10 units
area:
 600 square units
volume:
 1000 cubic units

Figure 10.1. The relationship between side length, surface area, and volume for similarly shaped solids—here, cubes. The small cube has a lot of outside relative to its inside; the large cube has a lot of inside relative to its outside.

question of whether it's tougher to heat or tougher to cool seems to depend as much as anything on how big one is. Of course, that's the same pattern we saw with metabolic rates—the small creature has denser capillary beds, shorter circulation times, and a higher rate of energy use relative to its volume. In fact, putting aside vigorous activity, it seems that the relatively better circulation of small mammals is a response to just such thermal imperatives, that oxygen is mainly needed for keeping up heat production. If the distance limitations of diffusion or conduction were decisive, then the larger animal ought to need the better circulation. Those limitations, however, are unlikely to be the most immediately relevant ones. All mammals and birds are probably so large that either diffusion or conduction is insignificant over appreciable internal distances. What matters more in practice is an animal's size relative to the problems of heat production to maintain the temperature standard of its class. In that respect it's the smaller ones that have the greater problem.

The Role of Flowing Blood

Were each bird or mammal quite single minded about keeping body temperature as constant as physically possible whether at the skin or in the deepest of its viscera, then the story of convective heat transfer using the circulatory system would be simple and dull. But why should an animal give top priority to thermoregulation? We have to remind ourselves from time to time that the only real biological imperative is reproduction. To exaggerate only a little, we individuals are just our genes' devices to reproduce themselves, and anything that furthers such reproductive ends is all to the good. Keeping warm is a useful if fairly expensive activity, but pushing the practical limits of temperature constancy would be much more expensive. Hibernation and nightly torpor have already been mentioned, but other deviations are a lot closer to home. Unless you're reading this while working an exercise machine, your fingers are colder than your arms, which are colder than your brain and visceral organs. An average skin temperature of 33 or 34° C (91 or 93° F) feels about right, while the fingers are cooler still. The "normal" or core temperature of 37° C (98.6° F) is reached only about 2 centimeters beneath the surface. At night your temperature drops about 1° C. If you exercise vigorously it will rise to 39° C (102° F) or even 40° C (104° F); at rest the latter is normally considered a worrisome fever. We're not fanatic players at this game of keeping a constant body temperature.

In vigorous exercise not only does body (core) temperature rise, but the temperature becomes more nearly uniform. For any activity besides staying warm, organisms are grossly inefficient, and most of the extra energy consumed in the activity appears as internal heat. For a human or any even moderately large mammal, the problem shifts from one of staying warm to one of getting rid of excess heat. Recall that humans are especially good at doing aerobic exercise for long periods. Concomitant with an increase in our energy consumption to tenfold or more above resting level is nearly the same relative increase in heat production. That problem has been invoked to explain our hairlessness, viewing humans as hot climate runners with a recurring problem of heat dissipation. For example, your hands commonly get hot when running. Blood flow to them has increased, and they're being used as radiators. When body temperature goes up for almost any reason, a call for greater heat loss goes out from the brain. As a result, flowing blood is shifted to vessels just beneath the skin, and you get red all over.

While perhaps a bit peripheral to the core of the present story, the rest of the path of the heat ought to be described. As mentioned, it takes a lot of energy to convert liquid water to gaseous water, that is, to evaporate the stuff. Properly done, that heat can come from an organism, leaving it cooler as a result. That's what we do when we sweat, although any liquid perspiration remaining on the skin represents water that has yet to do its part in the cooling process. Vaporization of water therefore cools the most superficial part of the skin, and that permits conduction of heat from blood in the nearby vessels—conduction and a temperature gradient get back into the picture. Of course, the blood is delivering the water for vaporization as well as the heat. The same evaporative process happens in the lungs. Exhaled air is hot and fully laden with water vapor; inhaled air is usually cooler and not saturated with water. Some heat is lost by direct convection of cool air across warm skin, and some by radiation from warm to cool surroundings; however, by and large, we're sweaters.[4] Faced with excess body heat, large terrestrial mammals (including cattle and horses) mainly sweat, thus increasing heat loss through the skin, while somewhat smaller ones (such as dogs, sheep, and goats) pant, increasing heat loss through the lungs.[5]

Despite our impressive metabolic scope, however, we're bit players when it comes to using tricks involving circulatory adjustments for coping with internal and external thermal changes. That giant iguana of the Galapagos, mentioned as a diver in Chapter 5, undergoes a drastic change in circulatory pattern when it plunges from warm rocks into the cold waters of the Humboldt current. On shore, the iguana's heart beats rapidly, and blood flows copiously to the periphery. In water, the iguana's heart slows down, and circulation shifts so that heat loss through the skin is almost indistinguishable from what would occur were the animal to give up internal convection entirely. Of course, reptiles have more versatile (if less powerful) circulatory systems than we do. Furthermore, cold legs aren't likely to be devastating to an animal that swims mainly with its trunk musculature.

Even insects get worked up about convective heat transfer. Since a flying insect has particularly hard working flight muscles, a lot of heat builds up in the middle part, the thorax, which is the part to which the wings are attached. Some large insects are at least partly air cooled, pumping a one-way wind through the thorax in large tracheal pipes.

Bees, taking flight, greatly increase and shift the pattern of blood flow, with the abdominal heart pumping blood through the narrow waist, and then through the flight muscles. By continuity the heated blood returns to the abdomen, where it's directed just beneath the outer surface, with the entire abdomen working as a radiator.

Heat Exchangers

Circulatory heat transfer plays the same role as the cooling system in a water-cooled automobile. In the latter, liquid coolant absorbs the excess heat of combustion as it moves through channels (= capillaries) in the engine block; it's pumped by the water pump (= heart) into the radiator (= skin and lungs); it's cooled there by forced airflow (= breathing, wind, and movement); and it returns to the engine block for another load of heat. What we commonly call the radiator is really a device more properly and generally called a *heat exchanger*. In the automobile, that exchanger links the internal liquid coolant and the external air (Figure 10.2). Of course, the engine block forms the complementary exchanger, transferring heat from the burning fuel to the liquid coolant. Likewise, a refrigerator has two exchangers, one within the insulated chamber, which moves heat from the internal air to the pipes of refrigerant, and another on the back or bottom, which moves heat from refrigerant to the external air. An air conditioner is similarly arranged, with one exchanger inside and the other outside the house.

I want to talk in general terms about heat exchangers in order to provide a context for what must be some of the most clever devices anywhere in this wonderful menagerie of circulatory machinery. An exchanger usually involves two flowing fluids and a solid between. While keeping the fluids separate, the solid permits conduction of heat between them. In our technology, one fluid is usually inside some pipes and the other outside; the pipe walls forming the exchange surface are invariably metallic. As mentioned, metals are outstanding in thermal conductivity, the critical property for an exchange surface. For some reason (or perhaps as a pure accident) no metal is used for any structural material anywhere in a living organism. Natural heat exchangers therefore take a slightly different route to achieving effective exchange

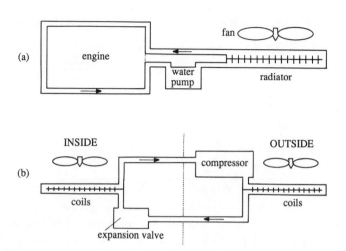

Figure 10.2. The arrangements for heat transfer in (a) an automobile engine and (b) a refrigerator or air conditioner.

surfaces—they inevitably use thin dividers of extremely great overall area. But, then, nature is good at making small things with lots of surface, such as tiny pipes and membranous walls. Beyond the consistent miniaturization, nature uses a variety of designs, either putting only one fluid inside discrete pipes, as in the technological examples, or keeping both fluids within pipes. As the working fluids she sometimes uses two liquids and sometimes a liquid and a gas.

We can recognize three basic geometrical arrangements for one of these heat exchangers. The first, called a *cross-flow* exchanger, is the familiar scheme used in automobile and air conditioner radiators that is illustrated schematically in Figure 10.2. In it the two fluids move at right angles (90°) to each other. Cross-flow exchangers are decent but dull devices, and I'll say no more about them here. In the second arrangement, the two fluids travel in parallel channels on opposite sides of the exchange surface, with both going in the same direction (Figure 10.3a). The engineers call this a *parallel flow* exchanger and the physiologists a *concurrent* exchanger. In the third arrangement, the two fluids travel alongside each other, but they go in opposite directions (Figure 10.3b). This one goes by the names of *counterflow* or *countercurrent* exchanger.

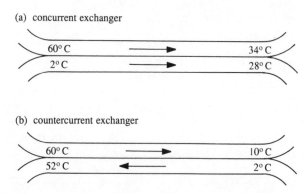

Figure 10.3. Heat exchanges arranged so (a) both flows go in the same direction or (b) the flows go in opposite directions. The temperature data were obtained with the exchanger shown in Figure 10.4.

While the difference in appearance between these latter two exchangers may be minimal, the functional difference is like night and day. I've put some numbers on the diagram—temperatures that can be achieved with a fairly primitive exchanger. They assume the simplest case—that the exchangers are decently insulated from the outside and that within each exchanger the two flows are equal. Under these conditions, the best the concurrent device can do is to give out fluids at a temperature halfway between the hot and cold inputs. In practice it won't quite reach even that mediocre optimum—the hot side won't quite drop halfway to the temperature of the cold input, and the cold side won't quite rise halfway to the temperature of the hot input. By contrast (and a little paradoxically at first glance), the countercurrent device can approach full transfer. The initially hot fluid can come out almost as cold as the cold input, and the initially cold fluid can come out almost as hot as the hot input. Now that's a good exchanger. It has a uniform temperature gradient along its length, rather than an unnecessarily large gradient at one end and an unusably small gradient at the other.

As far as I know, no concurrent exchanger exists in nature. These counterintuitively efficient countercurrent versions have been the inevitable choice whenever and wherever such exchangers have evolved.

Back in 1966, when I began teaching, I built several exchangers of the sort illustrated in Figure 10.4 for demonstration purposes. These could be run as either concurrent or countercurrent exchangers, depending on how the inputs and outputs were connected. Each consisted of 50 feet of ½-inch flexible copper water pipe run axially down 50 feet of 1-inch rubber automotive heater hose. I thus had two channels, inner and outer, separated by a good conductor, copper, with the whole thing isolated by a very poor conductor, rubber. The numbers on Figure 10.3 are about what I get from these exchangers, all of which are still used on occasion.

The Uses of Countercurrent Heat Exchangers

Up to this point I've talked about how circulating blood can move heat around, how it can take heat from an animal's core or from especially active muscles and dump it where it can be transferred to the environment. Strange as it sounds, nature seems to have paid a lot of attention to the opposite problem. How can a circulatory system be arranged so heat is not transferred? How can heat be kept in place when blood must move, carrying oxygen and other good things, over long distances within an organism? The solutions again involve heat exchangers—if blood is to move without moving much heat, then the heat has to be transferred from the blood elsewhere. It's for such a conservative purpose that nature makes her most radical and efficient exchangers.

The recurrent problem is perhaps best appreciated by considering the terrible dilemma of a gull paddling or wading around in a cold ocean. A gull is a proper warm-blooded bird, with an internal temperature of around 40° C (104° F). Its uninsulated legs stick down into near-freezing water, and the legs must be supplied by the gull's circulatory system. Insulating the legs adequately would certainly make them unsuitable as paddlers or walkers, but shutting blood flow would just as surely ruin them as legs. It looks as if the bird can't avoid sending hot blood down and getting cold blood back, thereby expending a vast metabolic effort in a futile but unavoidable attempt to warm the ocean.

What the gull does sounds on the face of it impossible. It attaches cold-blooded legs to a warm-blooded body, with both on the same circulatory system. A pair of countercurrent heat exchangers, just above

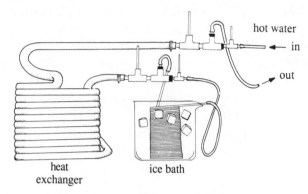

Figure 10.4. An exchanger, here shown with a countercurrent hook-up—a little impractical perhaps, but a good demonstration device.

the bare parts of the legs make the scheme work. Heat is exchanged between arterial blood going down into the legs and venous blood coming up from them. The arterial blood is thus pre-cooled *before* it gets to the bare parts, and the venous blood is prewarmed *before* it can chill the bird's innards, as shown in Figure 10.5. No free lunch need be assumed, just clever conservation. The same problem is encountered and the same solution is used by the flippers of porpoises and of other small marine mammals. Again an adequately insulated flipper would be a hydrodynamic monster; again a cold flipper is run on the same blood circuit as a warm body; again the trick is done with a countercurrent exchanger. A quick look back at Figure 10.3 will remind you of how one of these exchangers can work so well for such a task.

In addition to the countercurrent layout, the critical feature of such an exchanger is intimate contact between blood streams, arterial and venous, flowing in opposite directions. Mere contact between an artery and a vein ensures some heat exchange, but some of these countercurrent systems are much more elaborate and efficient. In a fancy exchanger, an artery branches into a bunch of smaller vessels, each a small fraction of a millimeter in diameter, that rejoin at the other end of the exchanger. Similarly, the corresponding vein branches, and then the branches rejoin. These tiny arteries and veins run in among each other in such a way that each artery is surrounded by veins and each vein by arteries, as shown in Figure 10.5. The structure was known long before

Figure 10.5. The arrangement of the countercurrent exchanger in the blood vessels of its legs with which a wading bird avoids heating the ocean. The cross section shows how the thicker-walled arteries and thinner-walled veins surround each other.

its function as an exchanger occurred to anyone;[6] it was called a *rete mirabile*, literally a *wonderful net*, for its appearance in cross section. We've usually viewed arteries branching and veins joining in the direction that blood flows, but no rule limits rejoining in arteries or re-branching in veins—what we commonly see is mere geometric convenience and economic design in the absense of any special circumstance such as heat exchange.

While these countercurrent heat conservers in legs and flippers almost surely have quite separate evolutionary origins, they play much the same role. A rather different role for a similar mechanism occurs in various large, rapidly swimming fish such as tuna and some sharks. We regard fish as cold-blooded animals, but again no strict rule bars fish from enjoying the benefits of nicely hot flesh and blood. During steady swimming, the dark muscles that mainly power such sustained activity produce, not surprisingly, a lot of heat. At the same time, equally un-surprisingly, they receive a plentiful supply of blood. The flowing blood would normally draw off most of the heat, transferring it through the gills, mainly, and through the skin, secondarily, to the surrounding water. Fish aren't at all well insulated, and in any case using gills to extract oxygen from water ensures very intimate thermal contact with the outside world. In these large and active fish, however, the arteries that run inward from beneath the skin to the dark muscles are surrounded by veins running outward from the muscles. Arterial blood is prewarmed; venous blood is prechilled—exactly reversing the heat transfer we saw in gull legs. It's still a case of a countercurrent exchanger used for conservation; again it keeps core hot and periphery cold with both on the same blood circuit.

To give yet another example, exercise causes a mammal, especially a large one, to get rather warm, as mentioned earlier. Large animals from warm, dry climates get particularly warm—core temperatures over 46° C (115° F) have been recorded for antelopes and gazelles. It seems important for an animal to avoid excessive hotheadedness, more important than avoiding temporary overheating elsewhere. One common cooling device makes special use of panting, in which rapid, shallow breathing cools the nasal and respiratory passages. Panting produces locally cool blood, which drains from the nasal passages through a pool in the head. Mammals that cool by panting commonly route the main artery to the brain through that pool. As it passes through, that artery is elaborately divided into a rete, and the flows run countercurrently. The scheme ensures that the brain will receive, above and beyond any other organ, the thermal benefit of panting.

To Conserve Heat or Not

To some extent the operation of each of these exchangers can be adjusted to fit the circumstances in which an animal finds itself. Blood can either be routed through the exchanger or most of it can be routed through vessels that bypass the exchanger. A duck normally loses almost no heat through its feet, but as the outside temperature drops below freezing an increasing amount of nonchilled blood is sent footward. Cold feet may be tolerable, truly frozen feet are something worth a little energy to prevent. A beaver has an exchanger at the base of its large, flat, and hairless tail; it's a nice thing if you swim in cold water. On a hot day, however, the well-furred beaver at pondside can keep cool by dangling its tail into the water, bypassing the countercurrent exchanger, and using the tail just like the radiator of a car. One common arrangement of exchanger in appendages has a central artery surrounded by veins. If more blood is directed through the artery, it expands enough to compress and collapse the veins. Thus blood has to return by other veins and doesn't pass through the exchanger, where it would take on heat from the artery.

We humans aren't equipped with any especially wonderful retia or top drawer countercurrent devices, but we do use the system in a somewhat less elaborate fashion. (One wonders, incidentally, whether the countercurrent exchangers in humans would have been recognized had the fancier ones elsewhere remained undiscovered. One more argument for biology in general and comparative physiology in particular!) Our arms don't have single central arteries and veins but come equipped with a variety of parallel channels for each. As *Gray's Anatomy* (Warwick and Williams, p. 698) puts it, "The veins of the upper limb can be divided into two sets, superficial and deep, which anastomose [interconnect] freely with each other. The superficial veins are immediately under the skin; the deep veins accompany the arteries." The description implies a central countercurrent system with a peripheral bypass; that's how the vessels are used. In the cold or at rest blood flows through the core of the arm; in heat or vigorous activity the exchange is bypassed. Try it, by looking at the degree of distention of your forearm veins when you're just a little cooler than you prefer. Then exercise vigorously with your legs for a few minutes and look again.

About half of us have yet another countercurrent exchanger. For reasons for which I've never seen a clear explanation, mammalian sperm do best at a temperature a bit lower than our accustomed 37°C.[7] Most mammals therefore put the testes in a special sac, the scrotum, which is hung loosely on the outside. Others, such as sloths, elephants, and whales, put the testes inside, but close to the outside body wall. In men the testes are kept about 3° C below the temperature of the abdomen. In part that's a matter of their cooler location, but it's also associated with the closely intercoiled arteries and veins of the spermatic cord. With that exchanger, the lower temperature can be achieved without limiting blood supply. In fact, it's been shown in dogs that blood is delivered to the testes precooled by just about 3°. Again, the system seems to be to some extent adjustable.

Once the basic scheme has been described, the temptation to prolixity is great—so many of these countercurrent systems exist and perform their exchanges in support of so many different functions. There's one in the wasp's waist that keeps heat in the flight muscles of the thorax when needed to speed preflight warm-up. Transfer of material across some placentas is aided by still another. Fish gills pass blood in one direction and water in the opposite to extract as much oxygen as possible from that water. Deep sea fish use them to aid secretion of gas originally dissolved in the ocean into their buoyancy devices, their swimbladders, and to avoid letting that gas dissolve back into blood and ocean. Small mammals, birds in general, and large mammals that live in deserts use them in their nasal air passages either to conserve heat or water or both. And on and on. Some uses involve circulatory systems, some don't. The stories would fill another book, one about nothing else besides heat and material exchangers in organisms.

At this point, someone might be wondering why, if these counter-current exchangers are so good, human technology hasn't done much with them. The main problems, I'd guess, is that elaborate branching arrays are awkward to manufacture. The one I made used a long serial exchanger rather than nature's version with an elaborate array of parallel pipes, but it's nearly as awkward to make and would be of practical interest only were heat conservation a matter of truly fabulous urgency. Someone gave me an exchanger designed for use in renal dialysis—blood goes in and out via tiny capillary tubes in one direction and the washing fluid passes over the tubes in the other. I once saw another designed for use as a simple portable device to humidify the air inhaled by people who've had tracheotomies. The moisture is provided by the previous exhalation, flowing in the opposite direction. Others undoubtedly exist in engineering technology, but the fact remains that they're far less common outside of organisms than are cross-flow exchangers.

Notes

1. Metals have far higher thermal conductivities than do nonmetals, although there's lots of variation metal-to-metal. Aluminum is about 400 times more conductive than water or muscle (ours or a roast). That, incidentally, is why aluminum cookware isn't prone to hot spots. Also, aluminum is four times more conductive than iron or steel, which is why a cast-iron pot would have to be four times as thick (and vastly heavier) to get the same temperature uniform-

ity. Even cast-iron or stainless steel is far better than the nonmetallic ceramics for stovetop use, especially on stoves with electric heating elements.

2. "Typically" hides a lot. When push has come to shove, nature has managed to design enzymes that work at temperatures as high as 80° C (175° F). These occur in so-called thermophilic bacteria that live in hot springs, but the latter are pretty special, even among bacteria. Thus it appears that one reason we run a fever when we have an infection is that the increased temperature hurts the bacteria worse than it hurts us.

3. Get out your handy calculator, look up the formulas in the dictionary, and try it. As Anna Russell entitled her biography, "I'm not making this up, you know."

4. The expression, "sweat like a pig," is especially inapt, for when it comes to sweating pigs are inept. A pig does sweat a little on the snout.

5. We're not well set up to pant, a special sort of shallow breathing. Done properly, as by dogs, it involves a particularly economical frequency of breathing. When we try, we blow off too much carbon dioxide, our blood gets too alkaline, and we feel dizzy or generally rotten.

6. Per F. Scholander, in the early 1950s, appears to have been the first to recognize the general mechanism and role of countercurrent exchangers. He was at once a brilliant experimentalist and a truly creative, imaginative thinker, one of the giants of a great generation of comparative physiologists.

7. Bird sperm don't seem bothered in the least even by the higher avian body temperatures. Strange.

11 *Tiny Vessels*

William Harvey, for all his fastidious dissections, measurements, and calculations, could only postulate the existence of the most important vessels of the vertebrate circulatory system, the capillaries. But Harvey's work, in the first half of the seventeenth century, antedated practical microscopes and microscopy, and the microcirculation is just that— micro. Perhaps I ought to emphasize for the reader comfortably uninitiated into the relevant arcana that microscopes and microscopy are not at all the same thing. Squint through an ordinary microscope at practically any bit of biological material and you'll see almost nothing. Light must be made to pass upward through very thin preparations in order to have much hope of visual revelation, and then it helps a great deal to provide optical contrast, whether just dark and light, or with contrasting colors. An elaborate contemporary technology concerns itself with how to preserve material without loss of structural integrity, how to embed it in wax or plastic or to freeze it, how to cut the stiffened result into slices a thousandth or ten-thousandth of an inch thick, and how to attach intense dyes to the slices so such thin pieces absorb enough light to be visible.

The first steps toward that technology were taken by the generation that immediately followed Harvey. And the combination of the experimental and conceptual legacy of Harvey, the new technology, and, I think it's fair to add, some very impressive people led to a period of explosive progress. Two of these people, both born as Harvey was publishing his most important work, figure in the present story. Marcello Malpighi (1628–1694), scrutinizing a host of different specimens, turned a primitive microscope on a frog's lung that had been inflated and then dried. He found the capillaries and immediately recognized them as the long-sought connection between arterial and venous systems. Shortly thereafter, Antony van Leeuwenhoek (1632–1723), also an extraordinarily innovative investigator, developed and used a series of awkward but competent microscopes. Among his other achievements, he verified and extended Malpighi's findings. In addition, he recognized and described the cells of the blood, first in frogs and then in other animals.

(Samuel Pepys provides a wonderful window into the center of gravity of all this activity, the intellectual world of Restoration London. He was inquisitive and acquisitive, a yuppie before his time; and his great diary is loads of fun to read. Not even microscopes escaped his scrutiny and comment. So as not to be too digressive, I've put the relevant entries at the end of the chapter.)

Why, by the way, frogs? Once more a particular animal turns out to be particularly suitable for a particular investigation. Frogs have at least three structures of sufficient normal thinness for observation of functional capillaries—their particular kind of lungs, the webbing between their toes, and their protrusible tongues. No less important, they (and amphibians in general) have especially wide capillaries and large blood cells, both about three or four times greater than those of mammals.

A Look At the Microcirculation

Earlier I slipped in the assertion that the smallest vessels are the most important. The statement shouldn't elevate eyebrows—if the circulation mainly moves material from exchange site to exchange site, then what happens at those sites must be significant. *Significant*, of course, isn't synonymous with *interesting*; each of us can name something admittedly the former but certainly not at all the latter. The smallest vessels, however, don't lack for good stories. The usual rules for flow are badly bent as blood cells quite literally squeeze through the capillaries in single file. Also, as we'll see, other peculiarities turn up and take on functional meaning. For all its importance, diffusion of dissolved gases

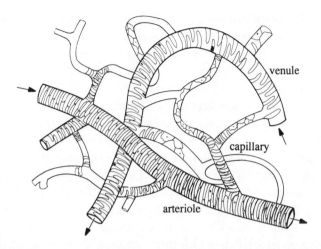

Figure 11.1. A somewhat impressionistic view of the different elements of the microcirculation.

is one of the more ordinary of the phenomena associated with the microcirculation.

To start with, we ought to view the relevant structures. Figure 11.1 illustrates the basic arrangement of capillaries, arterioles, and venules. The real layout isn't at all neat, tidy, and diagrammatic. Instead of a simple set of capillaries directly connecting the smallest arterioles and venules, something closer to a network appears, with lots of interconnections within the capillary bed. Still, that's not really unexpected for a system that readjusts its form in response to locally changing concentrations of chemicals that stimulate its growth. In addition, some arterioles and venules are connected directly, without any intervening capillaries (*arterio-venous shunts*), so blood can sometimes avoid passing the diffusive exchange sites altogether. Even the figure represents an idealization. One can't easily photograph a capillary bed except in occasional two-dimensional structures such as shaved rabbit ears, intestinal mesenteries, and those froggy items already mentioned. It's also probable that all these membranous tissues are substantially atypical. So we make reconstructions from drawings of adjacent slices of solid tissues, a procedure neither quick nor easy.

That the capillaries have the thinnest walls of any of these pipes makes obvious functional sense. For one thing, Laplace's law (Chapter

7) says that, being narrowest, they have least need for thick walls. Furthermore, the miserable dependence of diffusion on distance (Chapter 6) argues for thin walls from an entirely different starting point. One thin layer of cells makes up the capillary wall, and even these have rather leaky junctions where they abut.

The low internal diameter and wall thickness, however, aren't quite the extent of the uniqueness of capillaries. Among vessels only capillaries lack muscle cells in their walls. Arterioles and venules aren't always that much wider inside than are capillaries, but they have such cells, which by contraction can effect local changes in diameter. The results of such changes are major adjustments in the pattern of flow: to just which capillaries blood is directed, how much blood goes via capillaries versus how much goes directly to venules, even (because of the capillary interconnections) in which direction blood flows in some capillaries—all these are variables, undergoing both short and long term changes of substantial magnitude. Muscle cells make arterioles and venules fairly incompetent for diffusion, but the muscle layer matters little for heat transfer. Much use is made of the direct connections between arterioles and venules in your skin when you're hot, as mentioned in the last chapter. The point is that the microcirculatory bed as a whole is far from a static set of pipes conveying blood in response to pressure differences between large arteries and veins.

The capillaries of the walls of the alveoli of lungs have a substantially different arrangement from those anywhere else. Lungs, you'll recall, have an extraordinarily copious blood supply—they're on a low-pressure, low-resistance, high-flow circuit. The functional imperative, bringing a lot of blood into close proximity with the air in the alveoli, is straightforward. Capillaries ought to be close together so no area of alveolar wall is wasted. In addition, interconnections among capillaries ought to be useful for improving the uniformity of pressure and flow. A reasonable design might approach a double-walled covering on each alveolus, with the walls separated just enough to allow blood cells to pass easily between them.

What we have, as in Figure 11.2, is closer to that ideal than to conventional capillaries, something akin to a good quilted comforter with a double fabric wall and spacing struts atop a bed. Floor and roof of this blood chamber are about 8 micrometers (a three-thousandth of an inch) apart, slightly greater than the diameter of a systemic capillary.

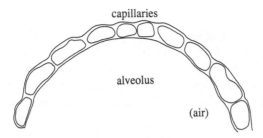

Figure 11.2. The way the capillaries of the lung surround the individual air chambers, the alveoli.

Thus, rather than thinking of flow through cylindrical pipes, it's more realistic to view blood as moving in what fluid mechanists call "flow between parallel flat plates" or "sheet flow," this name lying comfortably within our bedding analogy. Beside giving better coverage of the alveolar walls, less of a pressure difference is needed to drive sheet flow than tube flow if other things such as speed and diameter are the same.

Microcirculatory Flow

The basic rule that relates pressure and flow, the Hagen–Poiseuille equation (Chapter 5), presumes an ordinary sort of laminar flow in a cylindrical pipe. Ordinary means, among other things, that flow speed is maximum in the middle of the pipe, zero along the walls, and varies across the pipe in a manner termed "parabolic" (see Figure 5.5). While that rule works well for moderate and large blood vessels, it runs into difficulties for small ones. Mainly, the trouble traces to the red blood cells, which we noted make up nearly half of the volume of the flowing blood. When a vessel gets down into the size range of blood cells, then it no longer experiences blood as a homogeneous material. The vessel needn't be fully as small as a red blood cell—anomalies begin when it's ten or fifteen times larger. Even though the anomalies get most severe when pipes and cells are just about the same size, we're talking about a wider size range than that just of capillaries and their immediately associated arterioles and venules.

I really must emphasize that words such as *trouble* and *anomalies*, although literally reasonable, carry an excessive odor of constraints. We like rules such as that of Hagen and Poiseuille because they constitute simple and powerful descriptive and predictive tools. That the investigator is bedeviled by their failures to apply doesn't mean that the biological system works less well; it might even work better than anticipated. Perhaps we have here a fundamental difference between science and technology, between scientist and engineer. The engineer is in the business of design, and failure of the usual rule is at least a nuisance and perhaps much worse. The scientist's task is merely analytic, not synthetic, and a false prediction from the analytic rule that was expected to apply ought to occasion excitement and not disappointment. Previous plans may go awry, but we approach the realm of surprises, of places where nature can pull off wondrous, unanticipated tricks.

As mentioned much earlier, the viscosity—the resistance to flow—of mammalian blood is normally measured as about four times that of water. One way to look at the figure is to recognize that, for a given pipe leading downward from a given reservoir, blood will flow about four times less rapidly than water. For normal liquids, the size of the pipe has no effect on that four-to-one ratio. That's the case for blood unless the pipe is less than about a tenth of a millimeter (a $\frac{1}{250}$th of an inch) in diameter. The promised peculiarity begins to appear at that point. About 60 years ago it was found that as pipe diameter decreased below this size, the apparent viscosity of the blood began to drop. Not that blood flowed more easily through ever tinier pipes—that would really be strange—but it got increasingly watery, with a viscosity eventually falling to only about three times that of water. When the blood cells were removed and the plasma was tested alone, the anomalous decrease in viscosity in small pipes disappeared. That the phenomenon wasn't just some artifact of glass pipes in a laboratory became clearly evident when the pressure needed to force blood through the hind limb of an animal was measured. The more blood cells that were present (the higher the hematocrit), the more pressure was needed. The relative increase in pressure, however, was less for the limb, with a lot of tiny vessels, than for larger tubes.

That's very nice. Blood goes through small pipes more easily than we might expect, and this unusual ease (in small pipes relative to big ones) is associated with the presence of blood cells. So what's so special about blood cells? The overwhelming bulk of these, the red blood cells, are shaped like disks of hard candy or antacids, thinner in the middle than at the edges. In addition, they're rather flexible. Such a cell behaves strangely in flows where, as in Figure 11.3, the speed of flow varies with location across a pipe. For one thing, a cell rotates as it moves along. In a sense, the cell knows nothing about the overall flow since it's moving with the flow. (Its situation is that of a balloonist drifting in a steady wind, who can detect the wind only by noticing that the ground moves underneath.) The blood cell does experience a flow in the direction of the overall flow on the side nearer the pipe's axis, and a flow opposite that of the overall flow on the side nearer the pipe's wall. The combination, naturally, makes the cell rotate. In large pipes the speeds aren't much different on the two sides of the cell, so there's little rotation; but in small pipes the effect is a strong one for any cell not exactly in mid-pipe.

That rotation itself reduces the apparent viscosity of the flow. As you can easily envision, the speed difference causing rotation will be least when a cell is aligned lengthwise with the flow and greatest when the cell is broadside to the flow. Thus a cell quickly gets out of the broadside position but persists a little in the lengthwise one. Thus, on the average, a cell will spend more time aligned with the flow, where its viscosity-increasing effect is smallest, than broadside, where it makes greater mischief.

There's another part to the story of the flow of blood in narrow vessels, a part where the underlying physics is so devious that I'll not even attempt an intuitive explanation.[1] Since it has biological significance beyond mere viscosity reduction, however, it certainly deserves proper notice. In pipes only a little larger than the size of a blood cell, the cells preferentially travel closer to the center of the pipes than to the walls. That puts more of the cells in an area of flow where the local speed is more nearly uniform and thus where the fact that they aren't liquid is of least consequence. At the same time a relatively cell-free layer is formed near the pipe wall. The result is that viscosity is least where its effects are most pernicious, and for this second reason blood has a lower apparent viscosity when flowing through small vessels. This so-called *axial accumulation* also means that, on average, blood cells move through the circulatory system a little faster than does blood plasma.

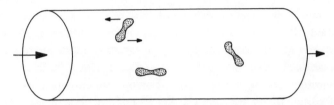

Figure 11.3. The forces on a red blood cell flowing off-axis down a small vessel tend to make it rotate.

The phenomenon is clearly a fine thing if the most demanding function of the system is provision of oxygen by those blood cells, since the hemoglobin can be used more often. After all, in the long run a faster vehicle can convey more passengers from one point to another.

Axial accumulation of blood cells has other consequences. The non-uniformity of the distribution of red cells allows branching vessels to be supplied with blood having a range of values of hematocrit. The possibility is clearly capitalized upon by kidneys. These organs work on the composition of blood plasma, and they process blood in amounts far greater than what would be needed just to supply themselves with oxygen. Red blood cells, therefore, must be a relative nuisance to the filtration apparatus of kidneys. In practice, the small arteries supplying the filtration apparatus leave the parental vessels at junctions flush with the walls of those parental vessels. The arrangement, called *skimming flow*, ensures that the small arteries will preferentially draw their contents from the cell-free layer near the wall.

The converse occurs as well. A daughter artery that protrudes into the middle of the parental vessel will be supplied with especially red cell-rich blood. The arrangement has been described for the branches of the main uterine artery in rats, although it almost certainly occurs elsewhere. The protrusion, quite reasonably if mechanical stability or cost of operation is of any consequence, isn't all that blatant—it's surrounded by what an airplane designer would call *fairing* and the anatomist calls an *arterial cushion*.

Finally, the fact that red blood cells prefer to travel in the middle of the road automatically ensures that smaller solid elements, the platelets, flow nearer the walls. Platelets are crucial in blood clotting, which is of course a way of plugging holes in the walls; being a little closer, on average, to the walls can't exactly be detrimental. Whether selected, designed, or fortuitous, the subtlety of these arrangements is awesome.

Everything said about flow so far applies to smallish vessels, but not necessarily to the capillaries themselves. Here there's no question of axial accumulation of cells since they barely fit through the pipes even going single file. Indeed, to equate capillary internal diameter and red blood cell external diameter is a little glib and misleading. Red blood cells are more nearly uniform in diameter than the capillaries through which they pass. The fit is sometimes easy, but what happens when red cells are larger than a capillary? August Krogh showed 70 years ago that deformation of the cells was not just possible, but normal; he even made movies of flow in the capillaries of the skin between the digits of frogs (frogs, once again!). The flat disks become more bell-shaped, with the apices of the bells downstream (Figure 11.4). The flexibility of the cells is crucial; if cells are artificially stiffened, then flow through capillaries is impaired. Conversely, a normal cell can be made to pass uninjured through a pore of less than half its undistorted diameter.

This tight fit, too, has subtle consequences. Back in Chapter 5 I mentioned that, because of the frequent passage of the red blood cells, flow is on the average faster near the walls and slower near the middle than it would be otherwise. Flow (as opposed to mere fluid) is in more intimate contact with the wall. That ought to expedite diffusive exchange, whether between blood cells and tissues or between plasma and tissues—an argument made in another context in Chapter 8. So by changing the physical character of flow, the mere presence of a lot of red blood cells can improve the rate of passage of material in and out of the plasma. A price, though, is paid for the close fit of cells to pipes. The relative decrease in viscosity as vessel size drops doesn't extend down to the size of capillaries. Significantly more pressure is needed to push blood through capillaries than one would calculate from the pressures needed for vessels just slightly larger. White blood cells, still larger but much less numerous, have even more trouble getting through capillaries. Fortunately white cells are independently motile, so blood pressure doesn't have to do the whole job. In practice, however, passage of a white blood cell amounts to a temporary blockage of a particular capillary—like a freight train on the passenger track.

A perhaps more immediate, if innocent, consequence appears if you

Figure 11.4. Distortion of red blood cells as they just manage to squeeze through a capillary.

measure the cellular fraction of the blood, the hematocrit, on blood squeezed from a finger tip—you may get a result either higher or lower than that from a sample drawn from a large vein. At the level of very small vessels, the concentration of red blood cells turns out to be re- markably variable.

Capillaries as Filters

Diffusion across the capillary wall suffices to transfer oxygen, carbon dioxide, simple sugars, and lots of other little molecules. Really large molecules, such as proteins, are mostly retained in the blood. A great diversity of molecules, however, are of intermediate sizes—all sorts of hormones and the like. For them, another arrangement comes into play, complementing diffusion in much the same way as does the bulk flow of circulation itself. Capillary walls ordinarily act as filters with very fine pores, pores too small for large molecules, but pores in quite the same sense as the holes in a sieve. The basic notion for how the scheme works was first proposed a century ago by Ernest Starling, who was mentioned earlier in connection with self-regulation of the heart. De- cent evidence for his idea wasn't obtained until around 1930, but it's held up since with relatively little modification.

The basis of the Starling hypothesis is the difference in pressure between arterial and venous ends of a capillary. In human skin, values of 40 and 15 mm Hg are typical for those ends. Pressures outside the capillaries are low, perhaps even slightly below atmospheric, so fluid with dissolved material ought to be forced out of the capillaries and into the surrounding tissue by the resulting pressure difference. Blood cells and large molecules are unaffected except that they are left a little more concentrated. This raises a question. What keeps all the circulating fluid from ultimately being filtered out, making (among other effects) your tissues swell? Starling's key insight was the realization that the presence of those large molecules provided a countervailing inward pressure of about 25 mm Hg. Thus at the arterial end of a capillary a net outward pressure (40 minus 25 mm Hg) pushes stuff out, while at the venous end a net inward pressure (25 minus 15 mm Hg) does the opposite. Figure 11.5 puts the arrangement in diagrammatic form.

Any respectable skeptic will wonder about that bit of deus ex machina, inward pressure from some large dissolved molecules. I could fall back on the magic word *osmosis*, but that would most likely just bring back bad memories of some confused or confusing teacher or textbook. It's better to look at a model that serves as a very close analogy (but not, to be honest, precisely an example of the real thing). If you take some yogurt and wrap it in cheesecloth, you can slowly and gently squeeze a lot of the water and dissolved material out of the proteinaceous gel. The result is a kind of low-fat, tangy cream cheese.[2] If you suspend the bundled cheesecloth and cheese in a container of water, a considerable amount of water will be slowly drawn back in. The protein essentially pulls the liquid back in, and that pull can be expressed as a pressure. It can even be measured, at least indirectly, as a pressure—one determines how much minimal outward squeeze would be needed to keep the cheese from pulling in water.

In short, there's a lot of flow in and out of capillary walls. Were the inflow and outflow exactly the same, little more would need to be said. As it turns out, the two are rarely quite the same. If you eat a lot of salty food and then slake your thirst, you may notice that a ring fits more tightly. Your blood volume hasn't changed much, but blood volume was maintained by a net flow of fluid from capillaries into the surrounding tissue that made you temporarily swell up just a little. If, and this is more serious, you are malnourished to the point of being substantially protein deficient, you get a characteristic sort of swelling or *edema*. In this case there's too little protein in the blood, and fluid cannot be properly pulled in at the venous ends of the capillaries.

The Lymphatic System

An imbalance between fluid pushed out and fluid drawn into the capillaries isn't just an occasional matter. The system normally has a slightly greater outflow than inflow, and yet another set of pipes and pumps deals with the matter. Everyone has heard of arteries and veins, but this third bit of plumbing, the lymphatic system, hasn't had particularly eloquent publicists. The system provides an additional route for returning fluid to the heart. The walls of its vessels are still thinner than those of veins, the pressures in its peripheral vessels are very low (only about 1 mm Hg), and it carries only around 1 gallon of fluid each day.

From almost all organs, then, very flimsy vessels arise, their initial segments starting as closed-ended but leaky tubes about twice the diameter of capillaries. Collapse of the tiny tubes is prevented, at least in part, by a crude looking array of filaments hooked to adjacent cells, much as some tents are tied to lots of points on an external frame. These vessels join (but also occasionally branch), eventually returning their contents to the venous system through two much larger vessels that join the latter just in front of the superior vena cava (Figure 11.6). Along the way many of the lymphatic vessels pass through lymph nodes, which are flattened, roundish bodies between 1 millimeter and 1 inch in diameter. These latter are intimately involved in the immune system, doing justice to which would require yet another book. They're also places where loose tumor cells may set up housekeeping and produce their progeny, part of another and far less pleasant story.

The fluid in the lymphatic vessels is called, unimaginatively, *lymph;* it's a little yellowish from the presence of the occasional red blood cell, but is otherwise clear. (There's nothing unfamiliar here—lymph is what accumulates in blisters.) It's basically the same stuff as extracellular (also called *interstitial*) fluid, which is basically the stuff pushed out of the blood through the capillary walls. Its composition is the same as blood plasma with but one exception: it has about half as much protein. That it has any protein might be a little surprising in view of what was said about capillary filtration. The filtration process, though, is a bit sloppy, and some large molecules do manage to leak through, to be

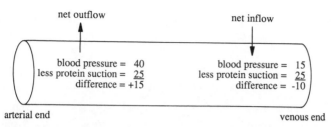

net outflow   net inflow

blood pressure = 40   blood pressure = 15
less protein suction = 25   less protein suction = 25
difference = +15   difference = -10

arterial end   venous end

Figure 11.5. A diagrammatic view of the way material moves out of and into capillaries according to the Starling hypothesis.

returned to the general circulation by the lymphatic system. In a day's time, half or more of the protein in blood plasma makes its way through the lymphatic system. Nonetheless, blood flows rapidly and lymph very slowly, so neither the level of protein in lymph nor this figure of 50 percent per day indicates more than a minor leakage. Still, over any significant time even that leakage would badly gum us up without some compensatory scheme.

The lymph vessels are generously provided with one-way valves, similar to those in veins. As in veins, some propulsion is accomplished by the squeezes provided by contractions of adjacent muscle and some by the irregular pushes and distortions caused by limb movements as a whole. The walls of lymph vessels, though, have their own muscle cells as well, especially near the valves. Thus the lymph vessels are best regarded as a lot of small, sequential, heartlike chambers arranged in a manner similar to the sequential hearts in some worms or in the abdomen of a lobster or shrimp.[3] The operation of the system seems mainly to be under local rather than central control. Accumulation of lymph upstream of a valve stretches the wall of the vessel, and the stretch stimulates rhythmic contraction of the muscle cells at about five shortenings per minute (in a sheep's leg, where it's been measured). Thus, like the intestine but unlike the venous system, the lymphatic system can actively propel its own contents. Ironically, it does very nicely what Harvey's predecessors erroneously thought that the other vessels could do.

Lymphatic systems are characteristic of vertebrates, whether fish, amphibians, reptiles, mammals, or birds—despite several statements I've encountered explaining why only mammals and birds had them or needed

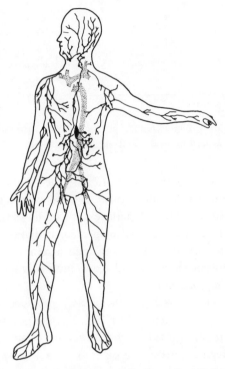

Figure 11.6. The layout of the larger vessels and the nodes of the lymphatic system.

them. Amphibians have a whole host of discrete "lymph hearts" that move the fluid and push it into the venous system. In some reptiles the lymphatic vessels are as conspicuous as the veins; reptiles typically have two pelvic lymph hearts that inject lymph into local veins. If you swing an arm rapidly and repeatedly, as when steadily using a hammer, the arm gets a bit edematous. Swinging a long tail in strong locomotory movements could get a reptile into the same difficulty; these lymph hearts ought to deal with the problem. Incidentally, an estimate of over 5 gallons has been made for the capacity of a lymph heart of the very large dinosaur, *Diplodocus*. The estimate isn't just an extrapolation based on body size, but it comes from imprints of the heart on the vertebrae. While lower vertebrates certainly have lymphatic systems, they do apparently lack lymph nodes. What occurs at the points corresponding to our nodes are usually just dilations of the vessels.

The necessity for lymphatic drainage of our extremities is evident from the results of blockage, either artificial or through the unkindness of some filarial worms. The result of the latter is a group of diseases occurring mainly in the tropics and characterized, among other things, by the most grotesque enlargements of the infected appendages. No parasitology book would consider itself complete without a picture of a (human) sufferer from the latter syndrome, called *elephantiasis*. Such pictures give a whole new meaning to the contemporary vernacular, "gross."

Parasitology is a subject of real intellectual fascination, but most of us were diverted to other areas by these images. What interests a biologist most about tropical parasitic diseases such as filariasis is that many of them don't really kill people, at least not right away. Humans were originally tropical and subtropical animals, and many of these parasites have had a long time to evolve a relationship with us. One can look at the situation from the point of view of the parasite and ask whether its fitness is really increased by killing the host. A biologist would suggest that the really clever (meaning well-evolved) parasite does better by taking over the host as a factory for producing more parasites. The host, to pursue the unpleasantness just a bit, is used as are the inmates of gulags or labor camps, which is what these filarial worms do.

* * *

I did promise a bit about Samuel Pepys and microscopy. Starting in 1662, someone named Richard Reeves was selling microscopes in London, so the aspiring biologist didn't really have to grind or cast lenses the way Robert Hooke and Antony van Leeuwenhoek did. Naturally, the entrepreneurial folk of the time had few scruples against selling instruments to anyone at all, no more than present salespeople who peddle encyclopedias to the illiterate and computers to the innumerate. A seventeenth-century microscope, however, was far from user-friendly, as is clear from the entry for August 14, the last with a specific comment on do-it-yourself microscopy. Still, Reeves' microscopes were beautiful, and Pepys' purchase undoubtedly graced the household, which mattered a lot to both Elizabeth and Samuel. He did keep up a vicarious interest, as the remaining entries attest.

Feb. 13, 1664. "Creed and I took Coach to Reeve's the perspective glass maker; and there did see very excellent Microscopes, which did discover a louse or mite or sand most perfectly and largely. Being sated with that we went away. . ."

July 25, 1664. "Thence to Mr. Reeves, it coming just now into my head to buy a Microscope—but he was not within."

July 26, 1664. "Toward [evening] I to Mr. Reeves to see a Microscope, he having been with me today morning, and there chose one which I will have."

Aug. 13, 1664. "There comes also Mr. Reeve with a Microscope . . . I did give him £5 10s, a great price; but a curious bauble it is . . . and [he says] he makes the best in the world. Thence home . . . to read a little in Dr. [Henry] Power's book of discovery by the Microscope."

Aug. 14, 1664. ". . .and then my wife and I with great pleasure, but with great difficulty before we could come to find the manner of seeing anything by my Microscope—at last did, with good content, although not so much as I expect when I come to understand it better.

Jan. 20, 1665. "To my bookseller's and there took home Hooke's book of Microscopy, a most excellent piece, and of which I am very proud." [The *Micrographia* of 1665, to which he almost certainly refers, must have been published only days before.]

Jan. 21, 1665. "Before I went to bed I sat up till two o'clock in my chamber reading Mr. Hooke's Microscopical observations, the most ingenious book that I ever read in my life."

It's easy to get a little wistful when contemplating a halcyon age during which cutting-edge scientific treatises were seized upon by the intellectual community at large. We've progressed to the point that the community is no longer interested, the treatise is no longer intelligible, and the commercial bookstores wouldn't think of keeping such items in stock.

Notes

1. A common textbook explanation, invoking either Bernoulli's principle or the Magnus effect, is most likely incorrect.

2. Hence the name *cheesecloth*. It was originally used to squeeze the water out of cheese curds.

3. This series of hearts is what you remove when you go to the trouble of "deveining" shrimp.

12 *Control*

By this point, it's certainly superfluous to dwell on the versatility of a well-evolved circulatory system. Upright or recumbent, even when submerged in water or weightless in a spacecraft, you get an adequate flow of blood to maintain your various organs. Be daring and jump out of bed sometime, and you'll feel only a momentary dizziness from any lag in the readjustment of flow to your head. Exercise vigorously, and a whole panoply of circulatory compensations occur—not only in heart rate and hence cardiac output, but also in the relative distribution of blood flow. Active muscles get more blood, inactive ones less, and the organs such as kidneys, liver, digestive gear, and so forth still less. These compensations, by the way, underly the likelihood of getting stomach cramps from heavy activity after a big meal. Mere anticipation of trouble—the late night ring of phone that triggers the parental gut-spasm—is enough to call forth all sorts of circulatory readjustments.

Beyond these changes responsive to one's external circumstances, circulatory systems undergo a range of alterations that maintain their own integrity as internal conditions change. Donate a pint of blood, and fluid quickly restores your blood volume by moving into the capillaries from between the cells. Drink a lot of beer, and two of the three effects reported by Macbeth's gatekeeper have direct circulatory relevance, ". . .three things does drink especially provoke . . . nose-painting, sleep, and urine." Burdened with the last, few of us pause to recall that the only route from gut to kidneys lies through the heart and blood vessels.

If you exercise for a while, your face gets hot and (if you're a fairly unpigmented kind of human) noticeably redder. As we noted two chapters back, excess internal heat has somehow led to an increase in peripheral blood flow that helps to dissipate that heat. The skin of the face, of course, played no role in making the extra heat. If (or, to put a finer edge on it, when) I cut myself shaving, I can more effectively stanch the flow by rinsing with cold water than with warm—the basic response to cool skin is a decrease in local blood flow. Quite conspicuously in the males of our species[1] certain familiar circumstances promote disproportionate blood flow to some spongy tissue within the penis. The stimulus isn't of necessity a local one, unless an unusually wide range of sensory capabilities are attributed to that organ.

The point of these examples is that elaborate systems of control are superimposed on all the fluid mechanical machinery of the circulatory system. Their main short-term functions are deciding how much blood should be moved and deciding how it should be apportioned among one's bits and pieces. On a longer time scale, control systems must maintain blood volume and composition in the face of both extraordinary accidents and ordinary changes of circumstance. On still longer time scales they adjust the structure of the vascular system itself, retuning variables such as the number and sizes of the vessels serving an animal's different parts.

As with so many of the subjects in this book, the complexity of the machinery contrasts sharply with the relative simplicity of the basic mechanisms. Thus an initial look at underlying mechanisms will keep us from getting lost in specific intricacies. As with the rules for flow, for pressure, for viscosity, and so forth, these mechanistic items aren't peculiar to circulatory systems or even to living organisms but are pervasive parts of both biology and technology. Indeed a very well-developed branch of engineering deals with this subject of control. Its principles transcend minor matters of whether connections are electronic, mechanical, hydraulic, chemical, or whatever. With such generality, this subject of "systems analysis" is directly applicable to living systems. So to set the present stage, it's worth digressing from circulation for a bit and exploring some notions of marvelously wide applicability.

The Basics of Control

Let me begin with what might strike you as a trivially mundane question. Why is the thermometer for a room or house located on an inside wall? Or, to put it another way, why doesn't a heating system use a device to sense the outside temperature, one that would tell it to turn on when the weather gets colder? After all, the thermometer on the inside wall can't distinguish between what the furnace or heater does and what the weather serves up.

Then there is a peculiar distinction. Some machines, while intricate, operate in a mindless, stupid manner. A spring scale will give the wrong reading if you use it on the moon—gravity is less there. An electric clock, of the ancient kind with motor and hands, won't work when moved to Europe from North America, even if the voltage is properly changed from 110 to 220. The clock will advance only 50 minutes in an hour since the electricity changes its polarity 100 times each second in Europe rather than the 120 times of North America.[2] By contrast, some seemingly simple machines have a kind of mechanical common sense. After a long period of nonuse, you may have to pour a little water down a drain to reestablish the liquid plug in the trap that keeps sewer smells where they belong. By contrast the water in the tank behind the toilet will remain at its earlier level. Somehow the tank has refilled itself just enough to offset any evaporation. Similarly, the thermostat keeps the house near your predetermined temperature despite wide changes of outside temperatures and whether or not windows are left open. In short, there are stupid machines and smart machines, and the distinction doesn't simply reflect levels of mechanical complication.

What, therefore, is the essential feature that makes a machine smart, whether household heating system or water level adjustor of a toilet tank? The answer is the following. *The smart machine can take a look at the result of its own activities, compare that result with some predetermined standard, and take action if the two aren't the same.* To do these things requires three crucial components. *First,* there must be some kind of sensor. The heating system has its thermostat, in which a drop in temperature leads to establishment of an electrical connection, and the toilet has a float, arranged so that a drop in water level causes it to move downward. *Second,* there must be some kind of comparator or integrator, in which comparison with the desired condition is made. For the heating system, it's the adjustment on the thermostat that sets the temperature at which the connection will be made, while for the toilet it's just the rod on the float, whose normal position can be crudely adjusted over a small range by bending it up or down. *Finally,* there must be the output device itself—furnace, water inlet valve, or other.

One more feature is necessary to make a machine smart. Say the machine senses a difference between the reference state (the setting of the thermostat, for instance) and the actual state (the temperature of the room)—what might it do? Two contrasting actions are at least theoretically possible. The machine might take action to *reduce* the difference between real and reference states. The furnace might turn on as the room got colder, or the toilet might admit water when the float went down. On the other hand, the machine might work to *increase* such a difference. But that's laughably inappropriate for our examples—a furnace that is activated by excess heat and a toilet tank that is stimulated to let in water by the addition of water. Thus the smart machine does the first—*it takes action to reduce the difference between reference (or desired) state and the actual situation.* (The stupid machine, of course, does neither.)

Why the thermostat is in the house should now be obvious. Arranged as just described, the system need care little about outside temperature, the thermal leakiness of the house, or heat from lights and appliances. Its internal weathermaking involves only the simplest of decisions—whether it's too cold or not too cold inside and thus whether to turn on or not to turn on. Similarly, the toilet tank needs no information on water pressure or evaporation rate; it just decides whether the water level is too low or not too low. One can make nice pictures of these machines such as those at the top of Figure 12.1, but such pictures, however attractive, mix up structure and function and typically end up doing justice to neither, a pet complaint of mine that came up quite a few chapters ago. The engineers do better, abstracting the

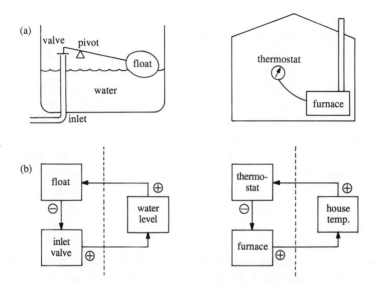

Figure 12.1. Toilet and heating system—negative feedback in your very own home: (a) the devices; (b) the corresponding functional diagrams.

functional essence in a block diagram of the kind given in the lower half of the figure. The arrows, by the way, refer to the transfer of information from one component to another, irrespective of the medium of transfer or of the encoding system. Information, of course, is the fundamental currency of control systems; it's no more (but no less) abstract a commodity than, for instance, energy.

With only a small additional item we can turn these abstract diagrams of control systems into quite useful analytic tools. In any proper control system at least one component (the comparator or integrator in the simplest case) has to *reverse* the sense of the information it receives before passing it on. An increase in the temperature of the house must direct a decrease in the activity of the furnace, and vice versa. A decrease in water level in the tank must signal an increase in admission of water through the inlet valve, and again vice versa. After all, reduction of a difference between two states is the essential task. The requisite reversal can be indicated by writing a minus sign next to the output arrow for the component that does the reversal, as I've done in the figure. Positive signs can be assigned to all other output arrows, mainly just to show that their logic has been examined.

This brings up two small but important points. First, every control system of the present kind must have at least one full loop of arrows, indicating that information can run around in a circle. That's where the term *feedback* comes in—a sample of the output is "fed back" to the input, closing the loop. Second, as noted already, for such a system to work the message must be reversed. This reversal is why we refer to these compensatory controls as *negative feedback* systems. The point is critical and all too commonly misunderstood. *Negative* doesn't imply less, or subtraction, or division—it refers to the *reversal* of the message that we indicated by the minus sign in the diagram. Put a little less abstractly, it refers to the grand function of *reducing* the *difference* between desired and actual states.

Just as it's awkward to describe *up* except by contrast with *down*, talking about negative feedback has a certain artificiality without mention of positive feedback. *In a positive feedback loop, action is taken to increase the difference between desired and actual states*, rather than to reduce that difference. For our examples disaster ensues—the toilet either empties or overflows, and the furnace either goes on full time or shuts down altogether. When compensatory control is called for, positive feedback is positively and pathologically perilous. That also goes for control in circulatory systems, to which we'll eventually return; but I want to emphasize what too often is missed—that positive feedback can play a positive role as well. It is, as we'll see in the next chapter, central to the mechanism by which blood clots and makes us self-sealing.

The minus signs in our feedback diagrams work just like the conventions for negatives in the English language or for multiplying positive and negative numbers. A single minus, one reversal, in a loop indicates that the system will exhibit negative feedback. Two minuses is equivalent to a single plus and hence positive feedback, just as a sentence with two negatives makes an affirmative statement. Three (or any odd number) of minuses means negative feedback—the diagrams are especially useful for such multiple reversing cases where verbal descriptions are always confusing. Only those in the legal profession put up with sentences having triple negatives; we scientists and engineers are simple folk and prefer to rely on diagrams. You can begin to appreciate the utility of such diagrams—make one with two minuses to represent a system with negative feedback and you have irrefutable evidence of a flaw in your analysis.

One more example will give you a homey reference for the use of feedback diagrams and for the different behavior of negative and positive feedback systems. I thought up this one after hearing a certain sad tale, but I've since found out that others have come up with the same model. Consider an electric blanket designed for double occupancy and equipped with dual controls, the situation shown in Figure 12.2a. Each side has a heat adjustment (whether the adjustment is itself part of a local feedback system is irrelevant here). Thus each person can reach down and take charge of his or her side. Each person, therefore, is sensor and integrator in a negative feedback loop. Normally, the wires from the controls come out at the middle of the foot of the bed—obviously the least obtrusive location. That, however, makes it the simplest of accidents to cross the wires, as in Figure 12.2b. After imagining the results, look at the diagram in the figure. One big loop has replaced the two separate ones. Person 1 controls person 2's blanket, and vice versa. It's awkward, but there's worse. The big loop has *two minus signs* and should exhibit positive feedback!

What happens? Before, we had compensatory control. If person 1 was too warm, that person would turn knob 1 down. Now, by contrast, if person 1 gets too warm, that person turns knob 2 down. Person 2 gets colder and so turns up knob 1. Person 1 is then even hotter and turns knob 2 still lower, making person 2 still colder—until whoever was initially too warm roasts beneath a blanket as hot as possible and the other person suffers frigidly under a completely nonenergized blanket. Give it any initial perturbation and the system, instead of compensating, maximally magnifies the perturbation. That's positive feedback.

Circulatory Controls

At the start, we need a notion that has been central to the world view of physiologists for over 100 years, one first stated by the great French physiologist, Claude Bernard (1813–1878). While it's hard to justify with any strict argument of evolutionary imperative, it has provided

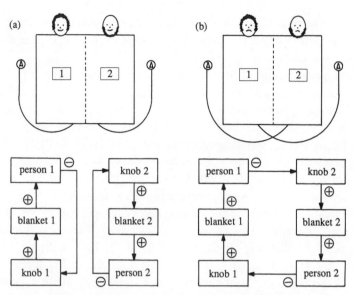

Figure 12.2. Electric blankets—(a) negative and (b) positive feedback even closer to home. Again the devices are above and the functional diagrams below.

invaluable and almost infallible guidance. Bernard distinguished between the external environment of an animal and what he called the "internal environment," that of most of its cells and which reflected the results of its activities. He argued that proper operation of our various systems required that that internal environment be a stable one, however the external one might fluctuate. Bernard didn't talk about negative feedback systems specifically,[3] but such devices are the clear and obvious way of doing what he pointed out needs to be done.

So which, among the elements of present interest, are kept stable? In the short term the principal items are blood composition, blood pressure, and, for mammals and birds, body temperature. These are regulated by adjusting other items such as cardiac output, the resistance of the vessels of the microcirculation, and the relative apportionment of blood to the various organs. Blood composition is a kind of umbrella quantity; we'll treat it as two components, with the respiratory gases first, and then with another umbrella covering everything else. We've recognized, then, four things that are regulated, and we'll have a brief look at each. Before we do that, however, I again have to remind the reader that quite a bit is being swept under the rug, perhaps even more than elsewhere in this book. For one thing, these four cases aren't fully independent systems, and they generate quite a tangle of interrelated feedback loops. For another, more sensors and output adjustments are involved than I'll mention. Beyond that, nonmammalian vertebrates will get no attention even though we know quite a lot about their regulatory arrangements, nor will invertebrates appear even though they clearly care about control as well. By "brain" I'll mean several chunks at the base or rear, the parts of our brains that handle our internal functions without commonly informing "us" about the details.

Oxygen and Carbon Dioxide Levels

Each of our cells, of course, normally uses oxygen and produces carbon dioxide. Not all the time, as it happens, though—energy can be liberated for short periods by reactions that don't use oxygen. These *anaerobic* reactions don't make carbon dioxide, either, but instead produce something called *lactic acid*. Still, oxygen is eventually needed to get rid of the accumulated lactic acid, turning it into water and carbon dioxide. Thus if you start to exercise a set of muscles, they will soon need an increased supply of oxygen. In part that's met locally by lowering the resistance of the blood vessels leading to those muscles by reducing the contractile tension of the muscles around the arterioles—*vasodilation* is the bit of jargon. That lowered resistance would ordinarily lead to lower arterial blood pressure, but let's briefly defer that problem. In part, the demand for extra oxygen is met by more distant changes. It is in these latter that the operation of a negative feedback system is most obvious.

If a person is shut in a chamber in which the concentration of oxygen drops and that of carbon dioxide increases, then both the person's pulse and rate of breathing gradually increase. If the experiment is done with a carbon dioxide absorber present so only the reduction in oxygen manifests itself, then those increases are delayed until the oxygen level gets quite a lot lower. The primary stimulus for the increases in breathing and heartbeat is clearly an increase in carbon dioxide. Much the same thing happens when muscular activity is increased, but the sensor in the feedback loop for adjustment proves no more directly associated with the active muscles themselves than is the household thermostat with the cold wave. Rather, the detectors are remote, and the same ones work no matter which muscles are active. They seem to be in the central nervous system somewhere—the exact location is hard to pinpoint since an increase in carbon dioxide makes just about any nerve generate more impulses.

It's a bit curious that the primary stimulus is carbon dioxide increase rather than oxygen decrease; functionally, of course, there's little difference, since production of the first is an inevitable result of use of the second. Perhaps the ease of designing a detector for something to which nerves are particularly sensitive made this the natural route for a designer devoid of any esthetic pride. The arrangement has a few curious consequences. If you suddenly go up to high altitude from sea level you initially breathe a little irregularly and feel strange. The lack of oxygen initially fails to stimulate a sufficient increase in your rate of breathing; you then overcompensate with rapid breathing, abnormally reducing the carbon dioxide level in the blood which in turn leads to a reduction in breathing rate, and so forth. If you breathe in and out of a plastic bag containing a carbon dioxide absorbant, you experience little of the discomfort of rebreathing or breathholding; you may even pass out before you feel any urgency about your deteriorating situation. (I certainly don't recommend that you try the procedure.)

More tempting, perhaps, is rapid and deep breathing before swimming underwater. At least at sea level you gain little in the way of additional oxygen loading since hemoglobin loads almost fully at normal alveolar concentrations. You do, however, blow off a lot of extra carbon dioxide, so you can swim further before breathing gets unavoidably urgent. Eventually, of course, carbon dioxide builds up and triggers real discomfort. People normally respond by surfacing and then breathe rapidly before their acid blood leads to heart trouble. Breathholding during vigorous activity isn't a good idea, but much worse is creating a situation in which overly competitive athletes are rewarded for pushing it to extremes.[4]

The system eventually does respond to lowered oxygen, and for that response the sensors are well known. Adjacent to each of a pair of large arteries in the neck is a tiny organ, weighing less than a ten-thousandth of an ounce. Each of these sends nerve impulses to the brain at rates dependent on the concentration of oxygen as well as carbon dioxide in the blood. Additional and similar sensors are located just off the aorta, immediately in front of the heart.

Thus the response to increased use of oxygen and increased production of carbon dioxide, as in exercise, involves a complex set of detectors, both local and central, and a complex set of actions. Casting them into a feedback diagram of the sort introduced earlier helps to keep track of what's going on (Figure 12.3); if we do it right, all loops ought to be negative ones. Local metabolic products (carbon dioxide and lactic acid, mainly) stimulate vasodilation in the active muscles. An increase in carbon dioxide in the blood stimulates an increase in pulse rate and thus cardiac output. As a backup, reduced oxygen does the same. Additionally, these changes in blood composition stimulate an increase in the depth and frequency of breathing.

An additional component outside the main feedback loops characterizes many of these systems. For instance, if you begin exercising, then the necessary increases in breathing and circulation start *before* the oxygen and carbon dioxide concentrations in the blood have changed. Both diffusion and circulation take time, and the body anticipates the lag. With the increased neural traffic to and from the muscles, the brain, of course, knows that you're up to something. The system therefore has an additional input, one that we might call a "change" detector as opposed to the "error" detecting sensors within the loops. Anticipatory control is always secondary, even though it goes into action earlier, and its role is reduced when the feedback loops get properly aroused. Perhaps I cast overly strong aspersions on putting temperature detectors outside a house. While they're a poor basis for a control system in the long run, they can certainly play a useful initial role. By using an anticipatory device, a house exposed to very sudden drops in temperature might be able to achieve a little closer control or get by with a slightly smaller furnace.

The longer a feedback system takes either to respond or to recognize that it has responded, the more important anticipatory control becomes. If you're thirsty, you drink water. If you kept drinking until the water was actually absorbed through the wall of the stomach, you'd have far exceeded that needed to replace the deficit underlying the thirst, and you'd then excrete a lot of water. In short, the system would ov-

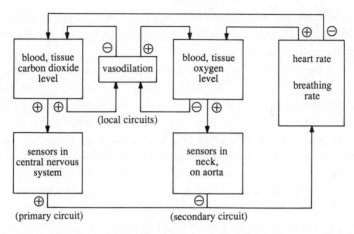

Figure 12.3. How we control the level of oxygen (and carbon dioxide) in the blood.

ershoot badly and perhaps oscillate. The same would happen with hunger, but similar anticipatory control by sensing things such as stomach distention compensates for the relatively long lag of the fundamental, nutritional loops. (But as a result you get only temporary relief by ingesting some nonnutritive stomach filler.)

Blood Volume

For reasons that at least to me are not self evident, we care a great deal about keeping the composition of the blood very closely regulated. The primary (but certainly not the only) organs involved in maintaining that constancy are the kidneys. Composition, as mentioned, is a composite of a very large number of separately variable chemicals and cellular elements. In the short term, the main shift is likely to involve increased concentration or dilution of all of them together. Thus if you lose a lot of water, they'll all increase; if you take on a lot of water, they'll all decrease. After all, if you dropped a quart of water into your gallon of blood, your hematocrit would drop from 45 percent to 36 percent. Quite reasonably, then, a major element in regulating the composition of the blood is control of blood volume.

It's useful to divide the water content of an animal among a series of what might be called "compartments." We are about 70 percent water and 30 percent solid, not atypical values for a large mammal. Looking only at that water, 64 percent is within our cells (including blood cells), 29 percent is in between the noncirculating cells ("interstitial water"), and 7 percent is blood plasma. That last is regulated with what seems fanatic commitment. Loss of circulating water is sensed by receptors at the base of the brain (and by pressure receptors, to be discussed shortly), and these trigger a whole set of events, cast into a feedback diagram in Figure 12.4. One result is thirst (with the anticipatory control mentioned earlier). Another is the secretion of a hormone by a gland at the base of the brain that reduces the output of the kidneys.

Yet another result of a decrease in plasma volume is a change of a particularly elegant sort in the resistance to blood flow through the capillaries. While both arterioles and venules have muscular walls, the two are under separate control. If the precapillary (arteriolar) muscles contract, then precapillary resistance rises. If the postcapillary (venular) ones contract, then postcapillary resistance rises. Now if loss of water triggers precapillary constriction, then the pressure in the capillaries will fall. By the Starling mechanism discussed in the last chapter, lower capillary pressure will force less fluid out into the interstitial space at the arterial ends of the capillaries. At the same time, more fluid will be drawn back in at the venous end. Thus blood volume will be restored at the short-term expense of the interstitial water. Still, the greater amount of interstitial than plasma water (29 versus 7 percent) means that a relatively small change in the former can produce a relatively large change in the latter.

Other elements of control also enter the picture. The short-term mechanism just described entails dilution of, among other things, the proteins of the plasma. Such dilution is intolerable in all but the short run since it would lead to excess water loss from the capillaries, which is just the opposite of what's needed. By mechanisms that I've not heard described specifically, the liver is stimulated by the presence of excessively dilute blood to release proteins such as albumin into the blood passing through it.

Superimposed on all this fanciness is an element of what might be called "passive control" or "autoregulation," an element so ordinary

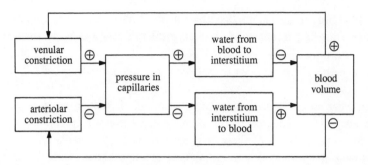

Figure 12.4. How we control the circulating volume of blood. Notice that one loop has three negative signs—proper negative feedback, but awkward to describe in words.

and direct that it's easily forgotten. If water passes into the blood-stream, then blood proteins are diluted. If blood proteins are diluted, then fluid is less effectively drawn into the blood at the venous ends of the capillaries. The net effect is a transfer of fluid from the plasma compartment to the interstitial compartment, reducing the magnitude of the initial perturbation. One might draw a feedback diagram for it, but that's a mild sophistry since no special machinery or pathways for information need be provided. Such passive effects are commonplace. You drive a car at a relatively steady speed in part because speeding up increases the car's drag and tends to slow it down, while slowing down decreases drag, permitting it to speed up again. The main trouble with these passive effects is that they can never be fully compensatory. Drag changes can't fully offset, for instance, the hills and dips in a road. By contrast, automatic speed control, using proper negative feedback control of fuel input, can keep speed regulated quite effectively.

Blood Pressure

Both the mechanical integrity of the circulatory system and proper exchange across capillary walls require that the pressure of the circulating fluid be kept within reasonable limits. The limits may not be as constraining as those for the levels of carbon dioxide, ions, or proteins; but without control mechanisms the activities of an animal would produce quite wild fluctuations. Vigorous muscular activity, as we've seen already, stimulates dramatic vasodilation in the affected muscles. If you used a lot of muscles at once and had no way to compensate, the resulting decrease in resistance would drastically lower your blood pressure. In actual practice, of course, pressure rises significantly in heavy exercise. Alternatively, exposure to cold stimulates peripheral vasoconstriction, which would automatically make blood pressure rise.

Average blood pressure is mainly determined by two factors—how hard the heart is pumping and the overall resistance of the microcirculatory vessels. Overall resistance, though, is an enormously complicated composite of what's happening in all the small vessels of all the organs. A change of circumstance may lead to an increase in the resistance of some, to a decrease in others, and leave still others unchanged; but it's in just such messy situations that negative feedback systems really shine. In a car, a good speed control system works over a range of slopes, of engine tunings, of altitudes, of loads, and so forth. The main trick, as should be obvious by this point, is to put the sensor where control needs to be most effective, where alterations in the controlled variable are least tolerable.

In mammalian circulatory systems, the main pressure sensors are located in the walls of the large arteries of the neck, not far from those carbon dioxide sensors mentioned earlier. Perhaps more significantly, they're not far from the brain, for whose functioning an adequate pressure is crucial and whose distance above the heart can vary. Pressure sensors, by the way, are really just stretch sensors in the walls of stretchy cylindrical vessels. I say "just" stretch sensors because these are used all over the place. If someone puts a book on your outstretched hand and arm, the appendage may twitch a little; but, even without watching, you normally keep your arm from sagging down to some lower level under the heavier load. Stretch receptors associated with your muscles detect the load and, by way of the central nervous system, trigger the muscles to work just enough harder to compensate for the extra weight. The operation of these fairly complex feedback systems is completely unintrusive upon our ordinary sense of control of our motions. Of course, other stretch receptors, such as those in the urinary bladder, are rather more blatant.

At any rate, these sensors in the arteries of the neck, and some others located elsewhere, play their expected role in the feedback loop. Any increase in pressure inside relative to pressure outside stretches the arterial walls and makes the sensors send more nerve impulses to the brain. In turn, the brain sends signals to the heart to change its operating characteristics, as in Figure 12.5. Thus an increase in blood pressure, other influences unchanged, will lead to a slowing of the heartbeat. There are a few complications, such as the facts that impulses are mainly sent (as you might guess) during systolic phase, and that activity of the muscles of the walls of the vessels naturally affects the stretch receptors within the walls.

The fastest way to louse up a feedback system is to fool the sensor. In my house only a wall separates the thermostat and the oven; thus, running the oven makes the house get cold. Sometimes, though, one can make use of the phenomenon. I've mounted a tiny heater under the thermostat, and the heater is plugged into a timer. On winter nights the heater goes on, and the temperature of the house is then regulated at a lower value. The fix was much cheaper than getting a thermostat with an internal timer, and it involved neither mechanical nor electric connection to the existing thermostat. It's possible to fool the pressure sensors similarly, but it's not usually a good thing. Pressing on them from the outside makes them think that the blood pressure is higher than in reality; if you press on just the right points of the neck you can induce fainting. Don't try it! I've read that accidental fainting was more common when men wore high, starched collars. I once witnessed a situation in which a lecturer canted her neck at a certain angle while looking at a projected slide and became briefly unconscious, apparently an occasional consequence of a minor vascular anomaly involving these sensors.

When talking about regulation of blood volume, I mentioned the possibility of autoregulation. For such a passive scheme to work, a system must have a certain intrinsic stability, which is not always the case. Quite the opposite behavior characterizes all too many situations. In these intrinsically unstable situations negative feedback becomes very crucial. One of the classic uses of a feedback arrangement involved James Watt's steam engines. To permit high-speed operation, he made the engine control its own input and output valves (for high- and low-pressure steam, respectively). The faster the engine ran, the more steam it

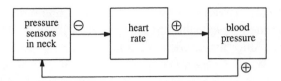

Figure 12.5. Our arrangement for controlling blood pressure, using as pressure sensors stretch receptors in some large vessels in the neck.

would let in, and the faster yet it would run. He therefore devised a governor, a device that responded to an increase in spin of the engine's shaft by reducing the input of steam. The result was imposition of a level of stability otherwise absent, which circumvented the intrinsic positive feedback and thus the natural instability of the engine.

Consider the regulation of pressure and flow in a circulatory system. If we had a set of rigid pipes, then an increase in arterial pressure would lead to a proportional increase in capillary pressure. With flexible pipes, however, the same increase causes stretch of the arterioles just upstream, a drop in their resistance, and therefore a disproportionate increase in capillary pressure. In practice, a bit of local negative feedback offsets the possibility. An increase in arterial pressure leads to local constriction by the muscles of the arterioles, and thus normal capillary pressure is restored. The scheme is useful when operating with the opposite kind of stimulus as well. Temporary shutdown of an artery, perhaps through constriction as a result of some ordinary postural or muscular event, results in arteriolar dilation. When flow through the artery is restored, there's a brief period of supernormal flow through the capillaries that it supplies.

All too many of us regulate our blood pressures at levels too high for our own best interests. Why the set point of the feedback system is altered is only occasionally clear, but one possible way is directly relevant to the present discussion. Many people with high blood pressure also have excessively stiff arteries, a condition known as *arteriosclerosis*. With stiffer arteries, a given pressure inside will less effectively stretch the receptors in the walls, so the system will "think" the pressure is lower than it actually is. Thus it will regulate at a higher pressure, just as my night-heated thermostat thinks the house is hotter and regulates at a lower temperature.

Unstiffening arteries isn't easy when the stiffening results from lipid build-up on their inner walls. On the other hand, a lot of chemicals are known that reduce blood pressure, although many of these tend to be unpleasant. My own experience with a fairly high dose of one suggests that much of the unpleasantness comes not from the reduction in pressure itself, but from achieving it by partially unhooking some of the connections in the control systems. The circulatory system then no longer responds appropriately to changes in circumstances of the sorts quite normal for active people.

Body Temperature

Despite the household examples involving temperature, I've deferred this one mainly because in thermoregulation the circulatory system is merely a vehicle as opposed to the target of the regulation. Also, it ought to be baldly admitted, the thermoregulatory system is a little more complex than the previous systems. Still, the individual items are closer to one's sense of reality, and most of the relevant items made their appearance in either Chapter 10 or 11. Here, for some of us at least, a diagram can do wonders—Figure 12.6 encapsulates most of the prose that follows.

First, temperatures of two different places are of importance—core and skin. Core is the inside of the body and the brain; skin is, well, skin. We tolerate wide changes in skin temperature, but only minimal changes in the core. Of course, each clearly influences the other, since heat flows from whichever is hotter to whichever is cooler. But the extent of the interaction varies, depending in particular on the degree to which the vessels of the skin are dilated. In particular, linkage between the two is heightened by a special kind of vasodilation—opening the arteriovenous shunts in the skin and thereby maximizing peripheral blood flow.

Second, what we might best call "extra" heat input can come from either of two quite distinct sources, internal or external—the muscular activity that accompanies exercise or the heat from sun, hot air, or some hot surface. Internal heat raises core temperature primarily and skin temperature secondarily; external heat raises skin temperature primar-

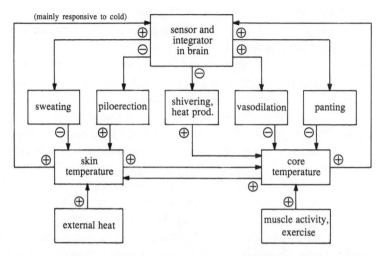

Figure 12.6. A more complex system—the mammalian scheme for regulation of body temperature. Notice that all possible complete loops are properly negative in sign, so the system is stable.

ily and core temperature secondarily. An external input of "cold" (in a proper physical sense) is just a reduction in heat and so needs no separate entry (although, as we'll see, the pathways for responding to hotness and coldness aren't quite the same).

Third, there are about five different kinds of output devices for this system, one device involving the rate of heat production and four adjusting the rate of heat loss. The relative importance of each varies with the particular circumstances in which an animal finds itself and, in addition, varies from animal to animal. We're especially good at sweating, but not at all effective at panting. Piloerection—fluffing up the fur—is a bad joke in humans, but of great use to more ordinarily furry mammals.[5] We begin to make use of increased heat production through shivering and nondirected increases in metabolic rate before we get as cold as most other mammals. In practice, though, we do a lot to avoid getting into situations where these responses are called for. Shivering, incidentally, is distinctly nontrivial. It's a kind of uncoordinated muscular activity with no other clear purpose than generation of heat. For short periods shivering can raise resting heat production as much as fivefold; for periods as long as an hour bouts of shivering can give a two- or threefold increase.

Microcirculatory adjustment, the final output, is here labeled "vasodilation" and connected with a negative arrow to core temperature. Alternatively, I might have labeled it "vasoconstriction" and put the negative sign on the previous arrow—each of these items automatically includes its opposite in such an essentially algebraic scheme. For instance, increased sweating decreases skin temperature while decreased sweating increases skin temperature. Either way, the sense of the information is reversed, the negative arrow is appropriate, and any awkwardness is merely linguistic. To keep things from getting just too complicated, I've omitted several microcirculatory loops that operate only under extreme circumstances. For instance, if skin is exposed to temperatures below about 10° C (50° F) for around 10 minutes, then the vasoconstriction ends and vasodilation begins; the shift is appropriate where there's danger of actual damage to the skin. I've also ignored the case where skin temperature might be higher than core temperature and vasodilation would therefore increase core temperature. It does happen—if, for example, you get into hot water.

The matter of anticipatory control came up earlier. The primary sensor in these thermoregulatory feedback loops is located at the base of the brain. If the climate were to change, it would be nearly the last place to know about the matter. The presence of temperature sensors in the skin is obvious to us all, since we quite literally feel their activity; but do they play a role in thermoregulation? The answer is slightly peculiar. They turn out to be much more important in the response to cold than in the response to heat. The trouble is that excess heat leads to an additional complication that reduces their practicality in creatures that make heavy use of sweating. When a person is exposed to a heat load, skin temperature changes in a direction opposite that of sweating rate—the more you sweat (really sweat, evaporating the water), the cooler is your skin. Since sweating cools the skin, almost as soon as sweating starts the sensors in the skin would send quite an inappropriate message, saying that all is well. That message would arrive long before any significant cooling of the core could happen, and it would thus turn off the cooling devices prematurely.

And, Finally . . .

These by no means exhaust the range of control systems, even short-term controls, operating either on or through the action of the circulatory system. For the somewhat longer term, one gets immediately into glands and hormones, and I want simply to remind the reader of the existence of such control systems rather than risk immersion in the intricacies of endocrinology. On the longest term (at least for individual organisms) we have control of the arrangement of the circulatory system itself. It isn't made once and for all, and we've already noted (Chapter 5) the sensitivity of the cells that line the vessels to the shear rate of flow across them.

Beyond that, as also mentioned earlier, the very pathways of at least the small vessels are subject to control in the same kind of feedback loops we've seen here. Tissues lacking sufficient blood supply release chemicals that stimulate the growth of blood vessels. Potentially active tissues, such as muscle, are especially avid players of this game; metabolically inactive tissues such as cartilage stay on the sidelines. The result is good news for such situations as readjustment after surgery; it's bad news when solid tumors get into the act and use the machinery to hook themselves to the host's circulatory system.

* * *

Some reader may have noticed that a certain dog hasn't barked. The word *homeostasis* was deliberately kept at arm's length, although it's been the standard battle cry for internal regulation in organisms and might even have headed the chapter. The truth is that I carry a small scar and a small evangelical impulse from something that happened in my formative years as a biology teacher. I once gave a lecture entitled "homeostasis" to a class with Knut Schmidt-Nielsen sitting in. After the lecture he asked me what this homeostasis thing really amounted to in specific terms. I mumbled a little and then looked up the original definition in Walter Cannon's book, *The Wisdom of the Body*:

> The coordinated physiological reactions which maintain most of the steady-states in the body are so complex, and are so peculiar to the living organism, that it has been suggested that a specific designation for these states be employed—*homeostasis*.

This is a can of worms of real metaphysical grandeur! Does the term refer to the state of constancy in organisms, or to the reactions by which relative constancy is achieved, or to some cosmic imperative toward constancy? Beyond that uncertainty is the statement's blithely biocentric view that such matters are solely the provenance of organisms. If that were the case, where would I have obtained all the nonbiological examples in this chapter? Organisms are quite wonderful and very special, but they share a lot with other complex systems.

The term may give less trouble as adjective than as noun, so one might safely call all the present devices "homeostatic mechanisms". Still, I'd rather stick with "negative feedback mechanisms" both to avoid a vague term and to aid communication with scientists not privy to biojargon. The general issues are that unnecessary proliferation of terminology does no one any good and that vaguely defined terms have always proven troublesome for science in the long run. We've escaped "phlogiston," the "luminiferous ether," the vitalist "entelechy;" with luck we'll evade "homeostasis" as well.

Notes

1. But not in all mammals. Quite a few get the requisite stiffening from a bone, the *os penis*. I'm told that those of raccoons find use as swizzle sticks and that ones over six feet in length are known from whales, useful conversational gambits.

2. Usually given as 50 and 60 hertz (Hz), respectively. One hertz is a full cycle of positive and negative waves, so it involves two changes of polarity.

3. Although they were certainly in widespread use in the latter part of the last century. Indeed, it's hard to imagine the industrial revolution without them. Otto Mayr has written a nice account of the history of feedback systems.

4. For reasons that I went into in an earlier book, swimming underwater is more effective than swimming on the surface. I was pleased to hear that the people who regulate the sport have banned staying underwater in backstroke races in response to discovery by coaches and swimmers of the competitive effectiveness of this dangerous practice.

5. I wonder if we carry some odd psychological scar as a result of having

lost fur sometime back in our evolution. Think of the popularity of fur coats, fuzzy toy animals, furry pets, and so forth. In a tactile room at a museum I visited, by far the most popular items were the samples of different furs; even for adults the urge to fondle them was hard to withstand. I once put my infant son on a fur rug, and he seemed overwhelmed by some great joyfulness.

13 *Leftovers*

A novel, history, or construction manual has an intrinsic linearity. Since reading progresses in the same linear fashion, shifting from reality to its written representation is a simple matter. The time axis may be stretched or compressed, but the speed of reading is by long tradition decoupled from that of real events. (Music, with a nearly inviolate velocity, carries the opposite tradition.) Loops and discontinuities in the time axis can be added for effect, but they're ultimately adventitious. By contrast, a book about a functional system lacks such a nice chronological baseline. So the writer concocts a progression and limits loops and arborizations to a few parentheticals and footnotes. The progression is contrived—and the material bent and twisted to fit. Nonetheless some relevant items persistently resist incorporation into any persuasive sequence. Here, then, are the last-but-not-leasts—or, implying less judgment or promise, the leftovers.

Blood Clotting

We are, of course, self-sealing. At least most of us are, and effective treatment is now available to restore the capability to those of us born without it. It's important in ways one rarely realizes. A tiny derangement of a molecule in one of Queen Victoria's ovaries, a molecule critical to inheritance of the ability to clot blood, changed the course of modern history. The defective version was passed on, and the actual consequent defect appeared in the son of the last czarina of Russia, Alexandra, who was daughter of Princess Alice, who in turn was the daughter of Victoria.[1] The scoundrel Rasputin could alleviate the consequences of the resulting hemophilia and thus gained authority over the czarina and, indirectly, the czar. The misrule attributable to his influence was a precipitant of the Russian Revolution. Considering the effect of that event on our century, we certainly should not be sanguine about blood clotting!

(Nor, conversely, can we blithely ignore the consequence of a cure of the ordinary sort for such inherited conditions. Exaggerating only a little for the sake of a tidy slogan, we might declare that if all hemophiliacs had children, then all children would eventually be hemophiliacs. The trouble in this case is that while hemophilia has been a devastating affliction for the afflicted individuals, female carriers turn out to be slightly more fertile than normal females. The bad gene is thus maintained instead of being selected against. Were hemophiliacs to live normal lives, the gene should actually increase in prevalence. I don't mean to be an alarmist, however—this isn't exactly the worst problem facing our species. Still, the lesson is that, even if manipulable, evolution remains inescapable. As the biopolemicist Garrett Hardin once pointed out, assiduous use of the rhythm method of birth control would lead to a rapid increase in the proportion of women who were arrhythmic.)

Blood clotting is a ticklish business even without genetic defects. Spring a leak, and the system must work dependably. Even more importantly, no plug must form at any other time or in any other place. A bit of clotted blood moving freely in the system is like a loose cannon on a wooden man-of-war—it can go almost anywhere, and none of the possibilities are attractive. Since arteries narrow down to arterioles and capillaries, some vessel somewhere will be plugged, depriving its branches of oxygen and other necessities. Matters are particularly serious if the clot lodges in an artery of such places as heart, brain, or lungs.

The process of clotting turns out to be exceedingly complicated, but the basic mechanisms are no longer especially controversial. In fact, two subprocesses contribute, first one and then the other. The two are complemented by a reaction quite separate from clotting itself—locally triggered vasoconstriction in damaged blood vessels. Of course none of these schemes is of much help when a really large vessel is severed.

Allusion was made to some tiny cellular elements in the blood when we talked (in Chapter 11) about axial accumulation. These *platelets* are two to three micrometers (around a ten-thousandth of an inch) in diameter, about a third of the diameter of red blood cells. Like the red cells, they lack nuclei and can't reproduce themselves. They're numerous, with about a quarter-million of them in each cubic millimeter of blood and something over a trillion in each of us, yet they account for less than 1 percent of the volume of the blood. These platelets play the central role in the first subprocess, which is essentially a quick aggregation of them.

Platelets don't normally adhere either to each other or to the layer of cells lining the circulatory pipes. Injury to a vessel, though, inevitably disrupts the continuity of this layer, exposing the underlying tissue. That tissue is made in part of the fibrous protein, collagen (which we encountered when talking about tough heart meat in Chapter 2 and about the mechanics of vessel walls in Chapter 7). Platelets adhere to the exposed collagen; furthermore, contact with collagen changes the character of the surface of the platelets through some chemical mediation whose details needn't concern us here. The adherent platelets are now sticky with respect to other passing platelets, which join and themselves get sticky—so a plug of platelets grows at the site of injury. The glob provides a quick but loose seal for a break.

You'll recognize positive feedback here. What stimulates platelets to get sticky, besides collagen, is the presence of sticky platelets themselves. As we saw in the last chapter, messing with positive feedback is dangerous, quite literally like playing with fire. (Light a match to a prepared pile of wood and paper, and positive feedback does the rest, at least until the availability of fuel or oxygen limits acceleration of the process.) The plug isn't a tightly coherent one, however, and protrusion further inward extends it toward regions where blood flow is faster. The faster flow then provides sufficient shearing force to offset further aggregation of platelets, and chemical factors produced by a normal vessel wall prevent aggregation up and down the vessel from the uninjured area.

The second part of clotting is coagulation itself, conversion of the gel-like platelet plug into a proper feltwork of protein fibers. Again the process starts as collagen is exposed by a break in the cells lining the vessels. In coagulation over a dozen chemical factors are involved, which gave rise to an enormous amount of confusion until the basic sequential scheme was recognized in the 1960s. Each factor is converted from inactive to active form by the presence of the active form of the preceding one. The final step in the sequence is conversion of a dissolved protein (*fibrinogen*) to an insoluble, fibrous protein (*fibrin*), which is the basic stuff of clots.[2] The arrangement is commonly called a *cascade*, but I find the analogy with a series of waterfalls forced and a little misleading. We actually have a multistage amplifier, familiar at least to people who work with electronic circuits.

While amplification is clearly necessary to get a mechanically adequate clot from an initial defect on a surface, it verges on tickling the tail of a dragon. The lengthy sequence of reactions has been interpreted as a safety device that works by keeping the actual amplification at each stage relatively low and well-controlled. That, however, carries bad news and good news. The bad news is that all steps must work properly for clotting to happen at all. The good news is that the process can be interrupted at lots of points, and so anticoagulant chemicals are no big trick to make and administer if they're needed. The bad news with that good news is that lots of creatures that bite or sting us have gone in for the manufacture of anticoagulants to keep their meals moving.

(A clot must not last forever, and the injured tissue does eventually repair itself. To deal with these matters we have an anticlotting system, in which a circulating enzyme slowly chews away at the fibers of fibrin. We also normally make some anticoagulants, but their exact role is uncertain.)

The active form of at least one of the intermediate coagulation factors stimulates the activation of its own inactive form. Here, as with platelet aggregation, positive feedback operates—the more of the active form one has, the more one gets. Both are entirely positive applications and not some pathological perversions of negative feedback. Other biological cases of positive feedback are more common than is ordinarily appreciated. A mating frenzy by some marine worms uses the scheme— the most effective stimulus to the release of eggs and sperm is the presence of eggs or sperm in the local water, so as soon as anyone starts, everyone gets into the act; and the chance of fertilization must thereby be vastly improved. The gland that stimulates the surface cells of insects to make new surface material, the crucial first step in molting, is sensitive to its own hormone. Thus as soon as any of its cells start secreting, the gland gives its all, and a sharp pulse of hormone is delivered. All surface cells consequently work in synchrony, which is crucial for a successful molt. Population growth and economic growth show positive feedback—the more people or resources available, the faster the system can produce people or resources.

As we saw in the last chapter, negative feedback systems are used as compensatory controls, restoring some variable to a desired value if it's shifted away. If negative feedback is appropriate, then positive is clearly a disaster; but as ought to be clear from these examples, positive feedback has a wholly separate and useful role of its own. It's a "now" command, a synchronization device. When a perturbation occurs, the system swings to either an upper or lower extreme, depending on the perturbation and the design of the system.

Working on Blood

From the start we've tacitly assumed that the demand for blood by an organ reflects its demand for oxygen. Certainly the assumption presents no problem for muscle, and muscle makes up almost half the weight of most mammals whether big or small. It's no problem for brain, for bone, for cartilage, and for a variety of other components. For skin a little modification is needed because of skin's thermoregulatory function, since that demands an especially generous blood supply when things get hot. To complement that picture, we now turn to several organs in which normal blood flow greatly exceeds anything that might be explained by the demand for oxygen. These organs have one thing in common—they do something to the blood itself. We're mainly talking about the lungs, kidneys, and liver.

It's worth reminding ourselves of just how much blood flows through these organs. Table 13.1 gives the relevant data, with comparative figures for a few other corporeal components. I've given them in two forms—as the fraction of the output of the heart (half-heart, really) going to each, and as volume flow divided by the mass of the organ. The data assume a 155-pound (70-kg) sedentary male human; coincidentally, they can be read either as liters of blood per kilogram weight per minute or pints per pound per minute. Or, if you prefer, they give the number of times each minute that an organ processes its own weight of blood.

Lungs

Through mammalian and avian lungs, of course, the flow must be equal to the flow through every other organ put together; the design of the dual circuit circulation makes that mandatory. Lungs are lightweight organs, and our two together amount to less than 1.5 pounds (0.6 kg), so they get an enormous blood flow relative to the amount of tissue present. Lungs, of course, are on their own special circulation with their own heart, the right half-heart, and that circuit is peculiar in many ways—noncylindrical capillaries, relatively low pressure and resistance, and so forth. Quite enough has already been said about the role of lungs.

Kidneys

What may come as a real surprise is how much blood goes through the kidneys. They get a substantial fraction of the total cardiac output— between a fifth and a fourth—even though together they weigh less than 1 pound (0.3 kg). They process almost as much blood relative to their weight as do lungs, and they handle fully eight times the flow of muscles doing heavy exercise! Running a pair of kidneys isn't especially cheap. They consume oxygen at better than half the rate of the resting heart, which has a much more obviously power-demanding action. Their blood supply, however, is vastly greater than can be accounted for by even this big demand for oxygen—they may get 22 percent of the output of the heart, but they consume only 6 percent of the oxygen used by the body.

One ordinarily thinks of kidneys as simply devices to ship out excess water and the byproduct, urea, of excess protein. In fact, they're

Table 13.1. Relative blood flow through various organs.

	Fraction of total flow (%)	Flow/organ mass (*liter/kg.min; pt/lb.min*)
lungs	100	8.0
kidneys	22	4.0
liver	13	0.85
brain	14	0.55
heart muscle	5	0.8
other muscle	18	0.03
(*in exercise*	75	*0.55*)
skin	4	0.08
(*maximum vasodilation*)	—	*1.2*)

crucial in fine-tuning the composition of the blood, ensuring that we maintain just the right concentrations of sodium, potassium, calcium, magnesium, chloride, and so on—the list is a lengthy one. They're arranged as a set of sequential processors. First, the smallish molecules are squeezed through a filter, removing them from the blood. This so-called ultrafiltrate lacks only blood cells and protein molecules; the process isn't fundamentally different from what happens by the Starling mechanism at the arterial end of capillaries (Chapter 11). The ultrafiltrate then passes to the pipes of the next stage, in which the good stuff is reabsorbed by a lining of cells and transferred back into the blood. Bad stuff or excess is left behind. What drives the filter, of course, is blood pressure; thus, a kidney uses the heart as a hydraulic engine. Reabsorption and other processes,[3] not filtration, use most of the oxygen extracted by the kidneys from the blood.

The amount of material ultrafiltered by a kidney is truly impressive—180 liters (50 gallons) per day is the figure usually cited for humans. So the urine that's finally excreted rarely amounts to even 1 percent of the original filtrate, which emphasizes the importance of reabsorption. Still, it should be pointed out that even that great filtration rate accounts for only about 20 percent of the blood that moves past the filter. The arrangement, again, is essentially capillarylike, with blood flowing parallel to a filter rather than butting up against a sieve.

The pressure available is a little higher than that in ordinary capillaries—typically 55 rather than 30–40 mm Hg. This extra pressure is obtained by supplying blood to the filtration apparatus through extrawide arterioles. (Recall that the largest pressure drop in the whole circulation normally happens in the arterioles.)

Blindly filtering out all but cells and the largest molecules and then reabsorbing most of the filtrate may seem a curious arrangement, but it has an interesting advantage. The system must only identify the good stuff, not the bad stuff. An animal can thus excrete a substance not previously encountered, either personally or in its evolutionary history. Not every vertebrate kidney, however, works this way—in particular the kidneys of some marine fish don't use a filtration stage but only secrete and reabsorb specific substances. This makes me wonder about whether they're especially vulnerable to environmental insults, such as the kind of chemical pollution that more versatile kidneys can take in stride? I know of no relevant investigations of the matter.

Liver

In Table 13.1 the liver doesn't look quite so impressive. That's mainly because it's one of the biggest of the internal organs, about five times the weight of the heart. If it got blood at the relative rate of kidneys and lungs none would be left to go anywhere else. In fact it deals with a lot of blood: it makes up only about 2 percent of the body's mass but receives 13 percent of the heart's output. What, then, does all that liver do with all that blood? Its cells are monotonously ordinary and homogeneous, and its overall structure gives little direct hint of function. The ancients gave it a central role in manufacturing blood, which was not a bad guess considering its rich supply of vessels and generally bloody appearance. A physiology textbook usually lacks a specific chapter on the liver that parallels the one on the kidneys and excretion and the one on lungs and respiration. It's not that liver function is especially obscure or mysterious, it's just that the notion of "function," singular, gets one off on the wrong foot. A list is about the only possible approach.

The liver acts as a blood filter, removing and detoxifying a variety of substances. It acts as a storage organ, particularly for sugars, but also for various vitamins and other molecules for which a ready reserve is a handy thing. It's the site of synthesis of a host of blood proteins, including most of the clotting factors mentioned earlier, as well as where cholesterol is made. The urea excreted by the kidneys is actually produced (ultimately from proteins) in the liver, as the latter scavenges the blood for the more seriously toxic ammonia and breaks up other circulating and dietary proteins. The liver is also an endocrine organ, producing several hormones. Not only is the liver heavily involved in production and storage, but it's the major controller of the concentrations of a whole bunch of substances that move around in the blood. For blood protein, fat, and sugar, it plays a role analogous to that of the kidney for ions.

In addition to its connections with the circulatory system, a liver has an outlet to the digestive system—a set of pipes leads to the gall bladder and intestine. Using these pipes, it's a digestive gland and excretory organ. Bile, which it exports, contains substances that act as detergents to help break up fat globules in the intestine. The breakdown products of the hemoglobin from ex-red blood cells are shipped out as well.

Others

A few other organs work on blood. As noted earlier, red blood cells in mammals aren't fully competent cells since they lack nuclei, so they can't reproduce and have to be manufactured. That happens in *bone marrow*, mainly in the bones of the chest, the base of the skull, and the forelimbs. We have a lot of bone marrow, and it produces red blood cells at an enormous rate: the average red cell has a lifespan of only around 3 months, and we have about 25 trillion at any particular time. Red blood cells are dismantled in the bone marrow, lymph nodes, and liver, but mostly in the *spleen*, yet another abdominal organ. The spleen is also involved in production of certain white blood cells and is thus a lymphatic and immunological organ as well. None of the functions of the spleen is unique to it, which is why a person can get along fairly well without one. Before puberty we have yet another immunological and thus blood processing organ, the *thymus*, in the upper chest. Also, in a proper list of organs that deal with blood one ought to include those tiny organs attached to the arteries of the neck that just sit there and keep an eye on the concentration of oxygen in the blood.

Portal Systems

Up to this point blood has always been viewed as going from the heart through a set of capillaries and then back to the heart. Only in one case was a situation mentioned where two sets of capillaries were sequentially interposed in the heart-to-heart passage. That occurred in fish, where we noted that blood goes first through the gills and then through the other capillary beds. Of course *we* wouldn't do anything so primitive.

In fact we do, and it works very well. It didn't happen to fit in earlier, but it goes well with liver fresh in mind. If you digest a meal, most of the absorption takes place across the walls of the small intestine; most of the absorbed material passes immediately into the circulatory system. The result might be a great rise in the circulating concentrations of all those things absorbed, but in practice a clever bit of circuitry circumvents the circumstance. The general arrangement is called a *portal system*, this one specifically the *hepatic* (liverish) portal system. The liver turns out to receive not one but two blood supplies, as shown in Figure 13.1. One, bringing in only about 20 percent of the blood, is an artery of the ordinary sort, carrying oxygenated blood in the usual way. The other, supplying 80 percent, is, of all things, a vein. The vein is the same one that drained the intestine—instead of entering the inferior vena cava, it plunges into the liver. It's the promised portal vein, a vein that goes from one set of capillaries to another. It's a vein that's formed by the normal convergence of smaller vessels but which then diverges in a most unvenous fashion.

The utility of the arrangement isn't at all obscure. Instead of pouring directly into the main circulation, the products of digestion go first to a big storage, conversion, and regulatory organ. Outside the portal system, there's a little change in blood composition after digesting a meal but the change is pretty minimal.

We have a second portal system that's a lot less conspicuous. It

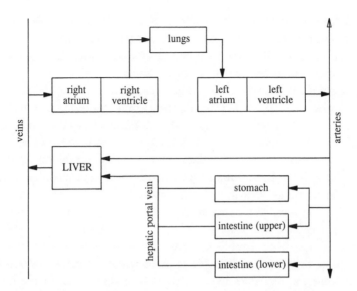

Figure 13.1. The circulatory connections of our liver and functionally adjacent organs, pointing out the way its portal system supplies it with venous blood from the intestines.

takes venous blood from a region at the base of the brain and delivers it to a gland beneath, called the *pituitary gland*. That region of the brain produces a number of substances, each of which stimulates the pituitary to release a corresponding hormone into the bloodstream. At least some of these substances go directly to the pituitary through this portal system. Utility? It certainly should take a lot less of some chemical to instigate a response if it's carried by blood directly to the target organ than if the chemical gets spread through the whole circulatory system. Flow in the opposite direction has been suggested as well; at least some fairly direct feedback path leads from the pituitary to the overlying brain. Since the system has a whole set of small portal vessels, flow in both directions might well happen simultaneously.

Lower vertebrates have a major portal system that we've given up. I say "given up" advisedly—mammals develop it as embryos and then lose it again, a vestigial trace of our reptilian ancestry. This "renal portal" system supplies the kidneys of fishes, amphibians, and reptiles with blood from their tails and hind legs. Kidney function takes a lot of blood, but this seems a strange way to get it. In fact the whole set-up is functionally peculiar from either of two viewpoints. First, it takes a decent hydrostatic pressure to work the initial filtration stage of a kidney, a point made just a short while back. That pressure demands arterial blood coming fresh from the heart, and these lower vertebrate kidneys do indeed receive the critical high-pressure blood supply. Why then, bring in a separate, low-pressure, low-oxygen blood supply to run the reabsorption stage of a kidney while sending the blood from the filters back to the heart without further use? Second, why should blood from something as critical to movement as the tail of a fish or the hind legs of a frog have to pass an additional capillary bed before returning to the heart? That's hard to reconcile with our notion that two capillary beds in series make trouble, the trouble whose avoidance rationalizes the double-heart system in octopus or us. In amphibians and reptiles blood from the rear does have an alternative route back to the heart, but that pathway takes it through the other portal system, the one supplying the liver, and so isn't really any more direct. Besides, what might the liver do that matters to blood coming mainly from muscles?

Perhaps we've not asked the right questions about this renal portal system. In science as in so much else, what one uncovers depends very much on what one goes seeking—nothing so biases an answer as the particular question asked. Dichotomizing perhaps a little glibly, people who study the structural arrangements of organisms, morphologists, have historically done so from two very different points of view. Structure may be viewed as an indicator of ancestry and structural comparisons may be used to postulate evolutionary lineages—that's the traditional approach of comparative anatomy. Structure may alternatively be viewed as an indicator of function, with the presumption that gross mismatch between structure and function will be so maladaptive as not to persist—functional morphology, to give the viewpoint a name. Harvey, unburdened by any notion of evolution and with his fine faith in a wise creator, was most certainly a functional morphologist; that may be why physiologists such as me (without at all disavowing evolution) find his approach so congenial. After Darwin's revolution, though, the comparative anatomists became thoroughly dominant, and most of the interest in matters such as renal portal systems has been theirs. They simply weren't primarily concerned with questions of function, and so functional questions have often remain unanswered simply because they haven't been asked.

In the present case, the basic question appears to be a functional one. Renal portal systems in amphibians and reptiles look functionally counterproductive compared with the arrangements of some fish and all birds and mammals. And they certainly aren't some minor vestigial blip on a morphological screen. Amphibians and reptiles are persistent, widespread, and numerous—successful animals, one has to admit. Why, then, hasn't the evolutionary process contrived a more direct heartward return for blood from behind? It would be a minor structural alteration, one that must happen by accident from time to time. The functional biologist suspects that the portal arrangement needs functional advantage to persist, but that we just haven't yet figured out what that advantage is. For a previous case of a persistent "primitive" condition, the incomplete ventricular separation in these same amphibians and reptiles, we had a decent functional explanation; for this one, we don't. The only moral I can draw is that we're only human, and matters of custom and culture influence our thinking much more than would certainly be optimal for scientific progress.

The Blood—Brain Barrier

The brain, as you'll note in Table 13.1, receives about 14 percent of the cardiac output of a person at rest; relative to its weight it's supplied with blood at the same rate as a hard-working muscle. For a person not doing strenuous muscular activity, the brain accounts for almost 20 percent of oxygen consumption. Thus it produces almost 20 percent of the heat generated by the body. As one might expect from all this metabolic activity, great arteries and veins connect the brain with the rest of the body.[4] One might expect that the nerves of the central nervous system, especially of the brain, would be quite literally bathed in blood.

As it happens, an intermediate fluid is interposed between blood and nerve cells, something called the *cerebrospinal fluid*. That's in addition to the normal interstitial fluid that makes up 10 or 15 percent of the volume of the brain (a figure not especially different from non-neural tissue.) Yet another fluid has appeared, and this one also circulates. In this case it's produced in one place, moves through passages, and is reabsorbed elsewhere. The rates and amounts aren't minute—a person has 4 fluid ounces (120 ml) of the fluid at a given time; about 24 ounces (720 ml) are produced each day. If exit of the cerebrospinal fluid is blocked, a condition called *hydrocephalus* results—literally "water on the brain." In a newborn infant the condition is marked by an abnormally large head.

The function of the cerebrospinal fluid seems to be at least in part mechanical. Brains are flimsy things, and the brain to some extent floats in the fluid, thereby avoiding certain mechanical insults. In addition, the cerebrospinal fluid is associated with a physiological phenomenon known as the *blood-brain barrier*. Small molecules diffuse from blood to brain much more slowly than they do between, say, blood and muscle. In addition, the composition of cerebrospinal fluid is quite different from that of blood plasma, so it can't be a simple ultrafiltrate as is interstitial fluid or kidney filtrate. Thus the cerebrospinal fluid is part of an additional regulatory system, another stage of buffering between the vagaries of the outside world and the world of the individual nerve cell. Claude Bernard might say "I told you so."

We've begun to take advantage of the barrier in designing pharmaceuticals. Passage is almost inevitably a nuisance for drugs whose primary target isn't the central nervous system. But it's often possible to tinker a little with the molecules or to find alternatives that retain the basic therapeutic effect without readily passing the barrier. A popular drug used for lowering blood pressure and for various cardiac conditions, the beta-adrenergic blocker propanolol, goes through fairly well; however, people taking it often get nightmares (but are commonly spared migraine headaches). Some newer drugs don't pass so readily and have fewer side effects. Similarly, most antihistamines pass through and cause drowsiness; a few have now been found that don't pass and still retain decent (if not quite the same) antihistaminic effectiveness.

Myoglobin

Talking about the blood–brain barrier involved asking about the fate of blood-borne substances after they leave the blood. Having brought up the subject, one can't pass on without mention of the further travels and travails of the most important of all blood-borne substances, oxygen. There's a nice story to be told, one that draws together a lot of loose ends of physiology, of biology in general, and of gastronomy.

We all know that meat comes dark and light, sometimes with both on the same animal. On chickens the breast and wing muscles are light, while the leg and thigh muscles are dark. By contrast, ducks have dark breast muscle as well. Fish have mostly but not all light muscle. Lengthwise strips of very dark muscle especially characterize steady swimmers such as mackerel, and some fishes such as tuna have fairly dark meat overall. We think of mollusks as a white-muscled bunch, but a very dark muscle runs the rasplike "radula" of the feeding equipment of some snaillike forms. What makes the difference between light and dark meat is the presence in the latter of a noncirculating form of hemoglobin.

The stuff is usually called *myoglobin*, although the name isn't quite as appropriate as it may have originally seemed. "Muscle hemoglobin" or "myohemoglobin" would be a little better, since it's clearly just a variety of hemoglobin, one used in muscle (= "myo") and kept out of the bloodstream. Our circulating hemoglobin, as mentioned in Chapter 9, is a tetrameric molecule, with four "heme" groups that bind oxygen. Our myoglobin is essentially just a monomer of the same—one heme group and an overall molecular size a fourth as large. If one examines enough animals, however, then one can find monomeric and dimeric hemoglobins as well as tetrameric myoglobins: they're really all the same jolly bunch of oxygen addicts. Wherever an animal has both hemoglobin and myoglobin, the crucial difference (beside whether or not the molecules circulate) is that the myoglobin has a substantially greater affinity for oxygen. It may not hold more oxygen, but it holds it more tightly. Thus if hemoglobin and myoglobin are brought into diffusive contact, oxygen spontaneously moves from the former to the latter.

What matters to us is, of course, function. The classical view of myoglobin was that it functioned as an oxygen storage facility. Muscles that needed a lot of oxygen for sustained activity kept some oxygen in reserve, bound to myoglobin. Undoubtedly that does happen, but the amount of oxygen that can be kept stored wouldn't take an animal very far. Then Per Scholander, whom I mentioned in connection with countercurrent exchangers, came through with another of his brilliant intuitive leaps. Scholander showed that oxygen diffused much faster through a solution of hemoglobin or myoglobin than it did through plain water. Not any dissolved gas was so affected—nitrogen, for instance, actually moved a little more slowly. Furthermore, the effect became more pronounced as he experimentally decreased the concentration of oxygen. In effect, the more oxygen a tissue uses up, the better this "facilitated diffusion" functions. In addition, the smaller the carrier molecule, the better the facilitation works; therefore, the monomeric muscle form, myoglobin, is more effective than the tetramer, hemoglobin. The whole thing makes good sense—myoglobin is found where facilitated diffusion should be useful, mainly in muscles that do sustained activity and consequently use oxygen at high rates.

Several loose ends are thus tidied up. First, the use of a tetrameric hemoglobin in the circulation was rationalized earlier; now the use of a monomeric version for the intracellular form is similarly explained. Second, a lot of the funny places that hemoglobins (here including myoglobins) occur now seem functionally reasonable. Animals that live in oxygen-poor habitats frequently have a lot of hemoglobin—tube worms that live on the bottoms of stagnant lakes, round worms living in sheep intestines, and lots of others. A storage function is unlikely to be the main use—a bigger gas tank is better only if you occasionally pass a well-supplied gas station. Conversely, facilitation of diffusion makes good sense if the concentration of oxygen around you is minimal and you still persist in using the stuff.[5] Third, the queerly sporadic appearance among organisms of hemoglobin as the circulating oxygen carrier is easier to understand. If facilitating diffusion is the primitive and more general role of hemoglobin, then putting the substance into blood represents a fairly minor alteration—it doesn't require a change in the underlying functional mechanism of reversible oxygen binding.

Perhaps of more immediate interest to most of us is how this functional view of myoglobin rationalizes the distribution of light and dark meat. As mentioned earlier, a muscle can be operated at least briefly without any use of oxygen at all. In the process, molecules of glucose (a simple sugar) are chopped in half, with the release of energy. Still, that won't do for long, since the half-glucoses (lactic acid molecules) accumulate and cause trouble; eventually oxygen is needed to deal with them. If a muscle is only used in brief episodes, however, then a capacious oxygen supply is unnecessary. It's these intermittently used muscles that lack myoglobin and have a light color. On the other hand, if a muscle is put to sustained use, then a good oxygen supply is mandatory and facilitated diffusion is a decided asset. So the muscles that do sustained work are dark when cooked or red when raw.

A chicken may be a bird, but it's only minimally a flying creature. Ducks and geese are real fliers, though, and the muscles of domesticated ones haven't changed all that much from their migratory forebears. They're proper birds, with very dark breast muscles for sustained flight. Oceanic fish commonly do both fast bursts and sustained swimming, and they have both sorts of muscles. A frog's legs, a squid's mantle, and a lobster's abdomen have light meat—they're used in episodic locomotion. We're looking at a very general and profound divergence in how an animal lives and how, concomitantly, it's arranged inside: cat versus dog, alligator versus porpoise—acceleration specialist versus velocity or distance specialist. Force matters more for acceleration than energy does. Thus acceleration is not demanding of oxygen supply—better to pack muscle with contractile elements and maximize its force than to fill it with equipment to use oxygen rapidly. Accelerators not only have less in the way of dark muscle, but they also get away with smaller hearts and smaller lungs or gills. Sustained velocity for the long chase, escape, or migration demands the opposite—the rate at which oxygen can be supplied and used rather than muscular force seems to impose the practical limits to performance. Weight training doesn't much help the marathon runner.

Notes

1. The defect leading to this particular kind of hemophilia is carried on the x chromosome. Thus females (with two such chromosomes in most animals) will be carriers and almost never have the overt disease while males who have the defect will inevitably display the disease.

2. In two books on my desk at the moment, pictures of fibrin meshes acknowledge the Gillette Company Research Institute. American males may find this amusing.

3. Some substances are specifically secreted by kidneys, not just passively filtered.

4. One is tempted to say "connect us with our bodies" since the brain contains what must be considered the person. I've long felt that the term *decapitation* was a misnomer; more appropriate would be *decorporation*—but perhaps this is not a topic for polite and refined discourse.

5. Fully oxygen-independent "anaerobic" organisms are well-known, but they're quite another story.

14 *And, Finally*

You've now come through what amounts to a textbook of circulatory physiology—a close approximation to what I'd have written had I been primarily interested in classroom use. The language has been less formal than in most texts, and the analogies and allusions have been a more diverse lot, but the only real omission has been a lot of jargon. As far as rigor and depth, you've put up with about what is ordinarily asked of undergraduate physiology students. I say these things to suggest to the reader who is not a scientist that science isn't all that formidable. One can join us at least intellectually and not be limited to looking in through some narrow window.

A Few Comments, Then, about Science

To me science is the most ordinary of human activities. It does impose a certain mental discipline, one based on rationality, logic, and skepticism. These, however, are precisely the mental habits that one calls upon when planning some household building project or trying to figure out what's wrong with your car. They're useful habits, well worth cultivating. I've taped an index card over a bench in my laboratory with a quotation from a book on doing science, by E. Bright Wilson, that says,

Often one's faith in cause and effect is put to a severe test by apparatus troubles; and the weak are likely to fall back on their ancestral inheritance of faith in gremlins and psychic phenomena. The persistent and materially minded scientist will, however, generally find the trouble by logical methods.

Whether wrestling with a problem at the desk or tinkering in the lab, there isn't really much more to it than that. Where trouble may come in is with aspects of the rationality and logic—our insistence on precise definitions, internal consistency, and quantitatively reasonable arguments. And with our pervasive skepticism—the distrust of what we disdainfully call "arm-waving" explanations and our categorical rejection of any kind of metaphysics or revealed truth. If you can't take such a hard dose of reality, then you'll have trouble with that building project, and the car's ailment will in all likelihood remain mysterious.

Perhaps the perceived impenetrability of science comes from another factor as well. As scientists we don't worship our forebears; but, as the present account certainly emphasizes, we don't forget their accomplishments either. While the decision by each generation not to reinvent the wheel underlies the wonderfully progressive character of science, it does make the endeavor very difficult to jump into on short notice. I treasure that progressive character; indeed, it may be why I'm a scientist and not, despite considerable interest, a historian. I do believe that the study of science has come much further since Aristotle than has the study of history since Thucydides, and I certainly mean no disparagement of historians by that assertion. The problem is that the progressive character requires that one establish a proper context in order to talk about almost any contemporary question or discovery. Trouble—impenetrability—comes when that context isn't set, something that takes time and effort and space and patience. So the nonscientist typically looks in through a keyhole and wonders, quite naturally, about what's really going on.

Yet another factor may underly the difficulty a nonscientist has in figuring anything out about the scientific process. That's a serious confusion about motives. A year or so ago a writer from our university's news service came around to inquire about my activities. I talked and she took notes, asking an occasional question. At the end, she asked me about the purpose of what I was doing, putting the query in a way that clearly implied technology or therapy. Even if I had a good positive answer to that, I think I'd not have used it, curmudgeon that I am. My actual response was to ask whether, had I been a historian or literary critic, she would have raised the question. Since she answered that she wouldn't, I requested the same indulgence.

After talking to several science writers, I think I'm beginning to understand their problem. The science is so hard to make interesting, because of the lack of proper context, that they ultimately have to hang a piece about science on something other than the science itself. Hence the "what good is it"—the handle for the reader and the imposed and inappropriate context for so many popular reports.

This misunderstanding about motives, though, can't be blamed entirely on science writers. Science has certainly generated technology (even as technology has empowered science, quite another story). What's hard to explain is a fundamentally paradoxical lesson from the history of science: the most effective long-term factor in science-derived technological progress is a strong scientific community that in the short term isn't concerned about that progress. My guess is that those scientists whose work has proven most portentous, such as Harvey, Darwin, and Mendel—I'll use biologists as examples—were driven mainly by a combination of curiosity and creativity. These latter are basic human characteristics, perhaps ones with which the evolutionary process long ago found favor. We are wired up in such a way that discovering something or creating something novel simply *feels* good. The arrangement may reflect useful characteristics for members of a species that was uncommonly dependent for success on innovation and versatility. In any case the motive, like the science itself, is thoroughly ordinary. Nonetheless it remains at least mildly paradoxical, at least as a recipe for practical progress. (Discovery has to feel good, for otherwise science is far too masochistic to attract all the people who do it. My favorite analogy for the process invokes a predilection for beating one's head against a wall and then, when the wall begins to crumble, picking another intact wall against which to beat.)

Beyond misunderstandings about motives, there are misconceptions about goals. Some of these have come up earlier; one in particular I think is worth mentioning at this point. The image of a scientific enterprise inevitably involves a lot of people working together to obtain something called *data*. One might reasonably infer that science is an organized activity devoted to the accumulation of data, and that science basically progresses through the accretion of data. However reasonable the inference, it's wrong. Few of us can do without data, but data are really only a means to an end and rather a nuisance at that. The end (and start) is a concept, while the data are mere steps toward deciding how much confidence should be accorded the concept. At some level everyone knows that, but it gets easily pushed aside in a world in which large sums of money and large numbers of people are engaged in obtaining "scientific data" without a clear conceptual question at issue. I don't mean that the data aren't useful, but that data are just a side issue with respect to basic scientific progress.

There is yet another peculiarity about doing science, and an especially bizarre one at that. In one sense we're intensely goal oriented—hence the disparaging remarks about "mere" data. In another sense, we're not particularly picky about the particular goal. After all, why complain if in looking for El Dorado you stumble upon Atlantis instead? Nothing I know of is so rooted in pure opportunism as is basic scientific research. Even in this era of hyperspecialization, the most productive procedure may not involve single-mindedly following a carefully planned sequence of experiments or assiduously mastering the standard methods of a particular field.

For instance, biology students almost inevitably come to graduate school with a commitment to some particular group of organisms. I've argued, to an extent just playing devil's advocate, that the commitment is needlessly constraining, that I've found it much easier to discover interesting things when I took living creatures as a whole as my permissible province.[1] But even so, the choice of organism, made near the start of an investigation, is just one small element among many of quite unfettered opportunism. Cases abound of a much wider opportunism, cases in which an entirely unanticipated goal was serendipitously reached. One rarely dwells on any midstream switch in writing up one's findings, but each of us learns early on that published papers make no pretense of being honest chronologies. That we accept any port in our perpetual storm doesn't mean that we get there in an expeditious fashion.

* * *

This book has been an experiment in context-setting, trying to establish a proper base for what you'll read and hear elsewhere about one small area of science. Newspapers and magazines are full of material on cardiovascular systems, and books abound on the proper care of one's own pumps and pipes. Since, as mentioned at the start, failure of circulation is what terminates most of us, this interest is entirely understandable; but the written material one encounters lacks adequate context, so the reader has no mental structure in which to place any of it, and so it goes, as we say, in one ear and out the other. Worse, when what one encounters is inaccurate, misleading, or downright nonsense, one has no way to apply the requisite skepticism. Little of what has been presented here was discovered in the past 5 or even 20 years—the basic science of circulatory systems hasn't had a lot of recent revolutions, despite news services and science writers.[2] If the odor of a textbook wafts forth—well, that's either the hazard of a teacher trying to write or (which I'd prefer to believe) because the present mission isn't all that different from that of a textbook. A textbook in science can have no better goal than to provide a context for reading the primary papers in its domain—its business is context-setting as well.

An analogy may again be useful. One might view the scientific enterprise as a very large balloon, with individual investigators and laboratories pushing out its membranous walls. At any time the nature of those walls and the devices for pushing on them loom large in determining interest, progress, or literature. The body of science, however, is the volume contained, not the interfacial membrane; the goals underlying any membrane-stretching activities cannot be understood without reference to that inside volume representing past efforts. Laplace's law (Chapter 7) as a metaphor for science!

Still, candor requires that I be explicit about my use of *a* context rather than *the* context two paragraphs ago. A context here means a particular biological context—I am, after all, a biologist, and more specifically a comparative physiologist and biomechanic. Bias lurks even within this area—you may have noticed a preference for physical over chemical explanations and detail. So that's the context I find most immediately relevant and about which I feel least incompetent to talk. It isn't the only context—one might look at our circulatory systems in the light of history, epidemiology, anthropology, sociology, medicine, and perhaps others.

And a Few Comments about Us

If cardiovascular diseases were rare in humans, you'd probably not have read this book; indeed, I probably wouldn't have written it in the first place. You've found very little explicit information about these diseases here, in part because I've not trained as a physician, in part because I find normal physiology much more fascinating (and a lot less depressing) than any kind of poking around among repair shops. Not only are cardiovascular diseases very common, they're a great deal more common than they were a century or so ago. When examined closely that last statement is profoundly disquieting.

One might dismiss an ostensible increase in any disease as reflecting no more than a change in diagnostic custom or sophistication. In this case, it doesn't look as if such a simple explanation will do. For instance, the specific sort of pain called *angina pectoris* has been associated with heart disease for several hundred years. Another explanation is that we are living longer, and heart disease is most commonly an affliction of old age. Teenagers may indeed have few heart attacks, but the bald statement that we are living longer may be quite misleading.

By a certain kind of averaging of data, we certainly are living longer. Life expectancy at birth has increased by 25 or 30 years in developed countries in the last century. What's a little misleading is that the figures usually given are for expectancy *at birth*. Birth and youth in general are vastly less hazardous experiences than they used to be; most of the contemporary change in life expectancy traces to the reduced risk of dying, mainly from infectious diseases, at fairly early ages. Once one has reached, say, 50, the remaining expectation of life has changed very much less. Our idea of how old one has to be to be an old person isn't very different from what it was 100 or so years ago. According to the U.S. Constitution, a person must be at least 35 years old to be president. When the constitution was written, life expectancy was 40 years.

Did the founders intend a gerontocracy, and ought we, in the interest of retaining the original intent, amend the constitution to require applicants for the job to be at least 65? Clearly not. While proportionately more of us are now old, *old* has not taken on a new meaning or shifted in any dramatic way.

What has changed, and it has been a watershed change, is the entire pattern of mortality. Death is now overwhelmingly associated with old age. Earlier, death was a relatively age-independent phenomenon. Your chance of making it through a year was roughly 98 percent, no matter how old you were (after your first year or two, anyway). In a world in which mortality was predominantly due to infectious disease what mattered more than your age was just what pathogen was running around at a particular time. Some years lots of people died; other years were really quite safe. Incidentally, when reading old literature, especially novels and diaries, it's helpful to bear this unfamiliar pattern of mortality in mind.

In short, while proportionately more people are old, old people are not much older. One does expect an increase in the incidence of cardiovascular disease where a population is on the average older, but the increase in our century is simply too great to be attributed entirely to that demographic transition. Furthermore the incidence of such disease is unmistakeably skewed toward Western-style industrialized or at least urbanized cultures.

Exactly what's going on is not easy to figure out. Comparisons with earlier populations are very difficult. One can look at old records, and the old parish records from England have yielded fascinating demographic information when analyzed. To a very limited extent, pathological studies can be done on skeletal remains. Comparisons with extant human populations—so-called cross-cultural comparisons—are far more practical, but they're more ambiguous as far as strict causation than is commonly appreciated. Only occasionally can genetic components be factored out. Thus Japanese have a lower incidence of cardiovascular disease than Americans. At the same time Americans of purely Japanese ancestry have the American and not the Japanese pattern, suggesting that the differences cannot be directly blamed on ancestry. Since Japanese cities are probably not polluted in some dramatically different fashion than are those of North America, this particular finger doesn't point strongly toward environmental insults. At least it doesn't point as strongly for cardiovascular disease as it does for the various cancers.

If the finger points anywhere, it points to personal life style; in particular it points to the components of one's diet, to the amount of fat we carry around, and to our levels of activity. Skinny, active, vegetarians come off well, even if many of us regard them as self-righteous fanatics best kept at arm's length. That does imply that personal intervention is an option, not an entirely unpalatable prospect. Further details, ranging from fairly certain to the worst mythology, are available at any library or bookstore—although one might wish for some hard-headed selection guide to exclude the excessively faddish, credulous, and theological. A decent biological perspective, perhaps even the one provided here, might be of some assistance in providing that critical crap detector.

One of my colleagues takes what I believe to be an enlightened view of primitive humans. He points out that they lived successfully for a few million years, changing their mode of life slowly enough during that time for natural selection to have kept them pretty well attuned to their circumstances. He suggests that it may not have been a bad life at all. Populations may have been kept in check by relatively benign factors such as high infant mortality, and there may have been very little infectious disease because of the low population density and infrequent interactions with anyone living afar. With agriculture, population growth, travel and trade, urbanization, and so forth, change became so rapid that genetic readjustment became insignificant, and we began to get out of biological "tune." I think he would have preferred to have lived 10,000 years ago at the latest; at least he makes a persuasive case that life for early humans may not have been all that nasty, brutish, and short.

We often imagine our ancestors as cave dwellers who lived by communal hunting of big animals. It's a heroic view, and certainly not unsupported by paleontological evidence. Nevertheless, it may be quantitatively misleading. For very basic ecological reasons big animals are usually rare, and only under unusual circumstances could they have provided a sufficiently steady food source for a community of humans. More likely, most early humans got most of their calories from plant material, food safer to obtain, easier to store, and more consistently available. If you want to put gender into the picture, imagine a culture in which the males (bigger, stronger, and faster) occasionally produced a fine kill, but the females (commonly pregnant or tied to nursing infants) provided the steadier supply of food.

The trouble with eating plants is that plants don't provide much energy for each unit of weight that one eats. Leaves are especially poor— animals that eat leaves need specialized digestive gear and even so have to eat for many hours each day. The gazelle grazes almost continuously; the lion needs only an occasional gorge of gazelle. Seeds, roots, tubers—storage organs of plants—are somewhat better, but even these are fairly dilute food sources once sufficiently hydrated to be palatable. I once calculated that it would take well over 1 gallon of cooked rice per day to keep a laborer energized, a calculation that followed demonstration by another colleague of how fast an oriental laborer normally eats a bowlful.

Thus in a world where food took a lot of effort to obtain and where periodic food shortages could be devastating, those big animals probably provided a rich treasure. The person who had a good appetite for pieces of large animal would be fitter, for that person might be more likely to make it through the next shortfall. Result—a taste for eating big game. This emphasis on the size of the animal eaten is deliberate. It gets us back to the material on heat transfer in Chapter 10. For a small mammal or bird, with a high surface-to-volume ratio, the major thermoregulatory problem is heat loss. It stores almost all of its body fat just beneath the skin, so the fat does double duty both as energy reserve and as insulation. The muscle, therefore, is fairly lean. (Even our domesticated chickens and rabbits largely retain the pattern, which is why skinned chicken is so good for weight watchers.) A large mammal cares less about minimizing heat loss; indeed, it may have trouble losing enough heat. It stores relatively less of its fat just beneath its skin, and it puts more within the muscles—well-marbled steaks come only from big animals.

Certainly, fat is the super deluxe food if you're interested in getting a lot of calories through a digestive system of limited capacity. *In palatable form*, a gram of fat yields about 9 kilocalories; a gram of carbohydrate with its inevitable water yields around 1 kilocalorie.[3] Is it unreasonable for nature to have wired us up to prefer eating the muscles of large animals? Perhaps much of our present trouble is that such muscle is too readily available, so we quite naturally eat too much of it. We base our rating of the quality of meat on its fattiness, and we raise animals considerably fattier than their wild relatives. To make matters worse, we also consume these fine animal fats in the form of dairy products such as cheese, again bending biology to our gastronomic predilections. A newborn mammal needs a lot of food in a big hurry; concomitantly, most milks have a lot of fat. By contrast adults wouldn't have had more than occasional access to dairy products before animal husbandry became prevalent.

From this biologist's perspective, then, it looks as if one problem is too easy access to foods we quite understandably prefer, but which we weren't designed to eat as a steady diet. Bits of plants with lots of unseparated fiber are not the happiest prospect, but they were probably our major sustenance over most of our existence as a species—before some of us got rich. Also most of our forebears were probably both skinny and quite active. Since thin people who exercise regularly have blood profiles that correlate with a lower incidence of heart disease, these latter factors may be as important as dietary composition per se.

Should you then become a skinny, active vegetarian? Here's another problem, one that's ordinary for a scientist but not quite so obvious if you don't do a little science. Everything said or implied about diet and activity has no more than statistical backing. It's about averages of data on populations, or it's about what happened to 400 out of 500 cases followed. Now no average is of much use unless one specifies how the individual items scatter about the average. To say that I average 1000 words a day doesn't mean that I write anything near 1000 words most days. Some days there's no new prose; on others 2000 or 3000 come forth. By contrast, to say that I average 65 m.p.h. on a highway when driving a car with an automatic speed control really does mean that I'm going very nearly 65 m.p.h. at any time. If high levels of blood cholesterol are associated with an increased incidence of heart disease, does that mean that you will be better off if you get your own level down? *In a very loosely statistical sense* you will be better off. Many people, however, have elevated cholesterol levels and never get into trouble, and lots of fairly corpulent and inactive people become respectably ancient.

The underlying problem of these averages takes on an awkward reality when we try to use them to make decisions for individuals. I used to ride a small motorcycle about 2 miles from home to the office each day. The accident statistics for motorcycles are truly alarming. Still, playing (as usual) devil's advocate, I argued that the people whom I knew that rode motorcycles were a particularly accident-prone bunch, and that I was a conservative college professor whose prospects weren't well predicted by statistical lumping with those wild folks. It may make sense to base national policy on population averages. Less national use of substances that make us prone to heart trouble will quite certainly improve the cardiovascular health of the country, but that doesn't guarantee that the same policy will work for any one particular individual.

Conversely, I would be the last to argue that the statistics ought to be ignored by the individual. Being in a very high-risk group as a result of prior trouble (but that's another statistical judgment!), I do emphatically practice what I only hesitantly preach. My wife was amused when I was described in a news report as "slight"; 10 years earlier "pudgy" would have put the matter kindly. And I exercise. And I live mainly on carbohydrates and fish. And I hope it's worth the trouble for me, not just for some statistical average; being average has never been a strong personal ambition. Anyway, as a cardiologist friend, Steve Roark, points out, the average male dies of heart disease. Moreover, I'd add, he does so without having written any books. I take the statistics seriously because I have no other information except perhaps my biologist's sense that what I'm doing is biologically reasonable. I fervently wish I did have better information, even if it just informs me that I am doing the right things to the right extent.

In practice we seek the services of physicians, who are, like the rest of us, a mixed collection as far as outlook, insight, and personality. Still, they do have a wealth of both personal and collective experience, and they can call on some impressive technology. On the other hand, they also carry a certain amount of bias and baggage. For one thing, most of them seem to lack the kind of horse-sense about statistics that practicing scientists acquire. For another, both the traditions of the profession and present attitudes of people toward doctors encourage tinkering and tampering. (I spent much of my youth in my father's small town pharmacy; it was clear at least then and there that physicians who overprescribed had more patients than ones who underprescribed.) Therapy, to both physician and patient, means either medication or surgery; talk about life style is just that—talk. I'm enormously impressed with the success of the rehabilitation clinic at which I'm a patient, but, as another physician I know put the matter, four out of five people who would surely benefit from its approach wouldn't possibly put up with it. Should the physician recommend it or place confidence in higher compliance with a less effective regime of pill taking—to put the question a little artificially as one of mutually exclusive alternatives?

Not only does the physician have to work within his or her guess as to what the patient can be expected to do, but there's another hazard, rarely recognized in our slightly physician-bashing culture.[4] Getting close to patients may be fine from the point of view of the patient; but in a life-but-too-often-otherwise specialty, such as cardiology, it presents special psychological hazards. If patients become family, then losing one is losing someone near and dear. As a frequent occurrence that's not something people tolerate well. Thus being a bit distant isn't entirely inappropriate behavior. By contrast, I thoroughly enjoy my contact with students, have become quite fond of many of them, and have been enriched by the encounters; but they typically survive the experience, and many go on to do wonderful things for which I like to presume some degree of responsibility.

Yet another level of biological reality may be hidden beneath these problems of disease and therapy. Back in Chapter 1 I mentioned that mammals, whether big or small, short-lived or long-lived, have a relatively constant value of a somewhat off-beat quantity—heartbeats per lifetime. The number, incidentally, is about 1 billion. It's amusing to take a simplistic view, imagining that in the absence of accidents or infections we live just until we've used up our allotted ration. It would suggest that exercise is beneficial because it reduces the resting heart rate and so more than makes up for the extra beats it incurs, that drugs that slow the beat prolong life, and so forth. As it happens, that view is quite indefensible.

"Heartbeats per lifetime," however, may reflect the existence of a more fundamental variable that has been called "metabolic time." Big animals have relatively lower metabolic rates, have slower hearts, breathe less often, live longer lives, and so forth. Small ones are the opposite. For reasons still shrouded in mystery, it looks very much as if an animal is in some sense used up after a certain time has elapsed, where time is measured on this size-related clock. Number of accumulated heartbeats is just one way to label the face of the clock. Still, the non-absolute character of the measure ought to be emphasized. For instance, birds live twice as long, on average, as do mammals, when lifespan is adjusted for body weight and thus measured in metabolic time. They do this despite their higher body temperatures, which should promote rather than discourage accumulation of chemical defects, especially in proteins.

This rather biblical notion of an animal's allotted time suggests difficulty in extending the lifespan of humans, perhaps the difficulty evident in the demographic shift discussed earlier. One might reasonably suppose that our traditional three-score-and-ten was about the point where a normal mammalian clock struck the witching hour. Again, the argument proves facile. As it happens, we're not at all typical of mammals when it comes to lifespan. We're substantially longer-lived than mice, elephants, and almost everything in between. On a mammalian scale of metabolic time we ought to live about 30 years, again barring accidents and other prematurely fatal events. So the three-score-and-ten is already something unusual, even if not completely bizarre or unprecedented. For our size and metabolic rate we're severely retarded, with late puberty, prolonged reproductive competence, a long postreproductive period, and delayed mortality. Perhaps it should come as no surprise that we are prone to geriatric difficulties.

Alternatively, one might take an evolutionist's approach and ask what advantage, in terms of our reproductive fitness, attaches to our particular lifespan. Why is it so long, and why do humans so often survive well beyond their years of reproductive competence? One might also ask why (in evolutionary rather than physiological terms) our lifespan isn't longer yet. The latter question may be the easier. If all members of a species lived forever, population stability would require a reproductive rate of zero. Among other things, that would preclude any evolutionary change. Thus it has been argued that the regular death of individuals contributes to the evolutionary progress of a lineage; however, all of this quickly gets even more complex, and no adequate context has been established for a decent discussion. (As I cheerfully admit to being hoist by my own petard.)

I do have one small suggestion, the Parthian shot of the inveterate biologist. Maybe something interesting might emerge from systematic study of the cardiovascular system of a mammal that's like a human in the particular way brought up here. I mean one that, like us, lives for an unusually long metabolic time, one that has a high number of heartbeats per lifetime. It should also be, like us, an omnivore or at least a meat-tolerant herbivore. At the same time it should be substantially smaller than a human, as of course most mammals are. This last condition would greatly facilitate investigation since it implies a shorter lifespan and more rapid progression or regression of the cardiovascular conditions of concern.

At the End, a Pitch for Science

I'd like to return to the issue of, as I put it earlier, "what good is it?" One can simply answer that it's fun, it's cheap, and it's harmless—each at least a half-truth. One can argue that science is the preeminent cultural event of our time, comparable to sculpture and philosophy in ancient Greece or orchestral music in nineteenth-century Europe. One can argue with peculiar logic that, while science often proves useful, useful science wouldn't get to first base without preexisting science that was done with no focus on foreseeable utility. Or one can point to our present predicament as we try to create a sustainable and humane global society for 6 or 10 billion humans, arguing that failure is certain without scientific progress.

One can make quite a different argument, of which I've not heard much talk, that a scientific establishment—scientists, neophytes, and acolytes working with decent facilities and support—is a most useful and at this point indispensable resource. During the Second World War, while we trained soldiers in a few weeks and military officers in a few months, the major technological developments were made by scientists and engineers who were recruited, educated, and professionally seasoned before the war. The requisite establishment could not have been created *de novo* in anything less than at least a generation. When the blizzard arrives, it's a little late to place an order for a municipal snow-plow. I don't at all believe that anything resembling that war is about to befall us; conversely, I'm firmly convinced that equivalent emergencies with strong scientific and technological imperatives are more likely now than ever before.

Consequently, it's imperative for us, the global community of humans, to maintain the resource. It can't be set up and then put in storage but must be maintained by continuous support. It's like a stable of horses, not a fleet of trucks. I'd argue, therefore, that support for a scientific establishment is socially worthwhile even if the results of its science are totally ignored. How nice that such a meritorious activity proves to be fun to do, fun to think about, and even fun to write about in an occasional book!

Notes

1. I really do. I've published papers using as material insects, the three major classes of mollusks, a red alga, sponges, mammals, and trees.

2. I once proposed something I egotistically called "Vogel's rule," a guide to determining whether a piece was written by a scientist or a professional science writer. The indicator is the word *breakthrough*. It's only rarely used by scientists, yet the science writers or their editors seem to be thoroughly smitten with affection for it.

3. The fat (butter, for instance) is taken straight; the carbohydrate (except for sugar, jam, etc.) must be taken with a lot of water. As served, rice and spaghetti have about 1.1 kcal/g, potatoes and corn about 0.8 kcal/g. The 9:4 ratio mentioned in nutrition guides is strictly true, but it's truly misleading. Carbohydrates are swell for dieting because they swell.

4. I'm as bad as anyone in this regard. A pathologist in our medical center has almost the same name as me, so I get lots of his phone calls. Every so often the following conversation happens: "Are you the real Dr. Vogel?" (Perhaps the question arises because I answer my own phone.) "Indeed I am a real doctor, not merely a physician."

Glossary

Where critical terminology ends and unnecessary jargon begins is very much a matter of the bias of surveyor, purveyor, or consumer. The present text has been pretty mercilessly purged of all but a few nonhousehold words that particularly facilitated matters. To save searching back through chapters rapidly receding from memory, I've collected terms that escaped excision here. In addition the list contains quite a number of items that appear nowhere in the text—these latter may help in dealing with less jargon-free written material and in direct communication with medical people. The definitions, I should emphasize, aren't totally general but focus on what's relevant to circulatory systems and this book. (Italics indicate entries to be found elsewhere in the glossary.)

Acetylcholine: Chemical that, released from the end of a nerve cell, affects a subsequent nerve cell or a muscle. Released by the nerve controlling the *vertebrate* heart, it slows the heartbeat.

Adrenalin: Chemical that, released locally by nerves or into the bloodstream, has a wide range of actions including speeding the heartbeat, increasing *blood pressure,* and constriction and dilation of microcirculatory vessels in various places. (Syn. epinephrin)

Aerobic: Referring to metabolic processes or to organisms as a whole that require an immediate supply of *oxygen* to function.

Alveoli: The tiny, saclike terminations of the *bronchial* pipes of the lungs, where gas exchange between air and blood takes place.

Amphibians: Class of *vertebrates* including frogs, salamanders, and mud puppies; derived from fishes, directly ancestral to reptiles.

Anaerobic: Referring to metabolic processes or to organisms as a whole for whose functioning no immediate supply of *oxygen* is needed.

Anemia: Any condition in which the blood contains a substantially subnormal concentration of the *oxygen* carrier, *hemoglobin*.

Aneurysm: An abnormal dilation or outward bulge in the wall of an *artery* or of the heart (occasionally "aneurism").

Angina pectoris: A constricting pain in the chest and secondary locations (left arm, jaw, etc.) of variable severity that reflects insufficient blood supply to the heart muscle.

Angiography: Diagnostic procedure involving injecting a substance that is opaque to x-rays into circulatory vessels and then viewing or photographing them with x-ray illumination.

Angioplasty: Insertion of a balloon-tipped fine tube into a blood vessel (commonly the large *coronary arteries*) and inflation of the balloon to reduce the degree of obstruction by deposits on the walls of the vessel.

Anoxic: Devoid of *oxygen*, usually referring to some habitat.

Aorta: The largest of the *arteries*, coming out of the left *ventricle* in higher *vertebrates* or the sole ventricle in lower *vertebrates*.

Aortic valve: The one-way *valve* between *ventricle* and *aorta* permitting flow from former to latter but preventing the opposite.

Arrhythmia: An irregular heartbeat of one sort or another.

Arteriovenous shunts: Direct connections between *arterioles* and *venules* without intervening *capillaries;* especially common in skin, which gets especially red when such shunts are open for blood flow.

Arterioles: The smallest *arteries*, but arteries still, with fairly stiff walls.

Arteriosclerosis: Hardening (stiffening, strictly) of arterial walls from any cause. See also *atherosclerosis*.

Artery: Vessel conveying blood away from the heart, including the *aorta* and the *arterioles*. Oxygenated blood flows in the arteries of the *systemic circuit* and deoxygenated blood in arteries of the *pulmonary circuit*.

Arthropods: One of the three major groups of really complex animals, the other two being *vertebrates* and *mollusks;* includes crustaceans, spiders, insects, and some other animals all of whom have legs with joints and external hard surfaces rather than bones.

Atherosclerosis: *Arteriosclerosis* due mainly to deposition of fatty material or the insides of large and medium-sized *arteries*.

Atrioventricular node (A-V node): A group of specialized cells on the right *atrium* near its junction with the *ventricles;* these cells are triggered by signals from the *sinoatrial node* to initiate the electrical events that will lead to ventricular contraction.

Atrium: One of two (one in fishes) thinly muscularized chambers of the heart that receives blood from the *veins* (sometimes through another, upstream chamber) via the *superior and inferior vena cavae* and pumps it into a *ventricle.*

Baroceptors: Sensory cells located in the walls of blood vessels whose output impulse rates vary with the *pressure* within the vessels.

Barometer: Any of a variety of instruments that measures atmospheric *pressure;* thus a particular sort of *manometer.*

Bernoulli's principle: The rule that, in the absence of *viscosity,* an increase in the speed of flow of a *fluid* will be offset by a decrease in the pressure of the fluid against a surface along which it is flowing. Of some relevance in heart *valves,* it otherwise has little applicability to circulatory systems because of the distinctly nonnegligible *viscosity* of blood. After Daniel Bernoulli, 1700–1782, Swiss mathematician.

Blood pressure: The *pressure* exerted by blood on the walls of the blood vessels; usually used to mean the pressure extremes in the large *arteries* when measured at the level (up and down) of the heart. The extremes are termed *systolic* and *diastolic* pressures.

Blood–brain barrier: A barrier restricting or slowing passage of many chemicals from the blood to the *cerebrospinal fluid* that bathes the central nervous system, mainly the brain.

Bohr shift: A decrease in the *oxygen*-combining avidity (not total capacity) of *hemoglobin* or other respiratory pigment in the presence of *carbon dioxide.* It facilitates unloading of oxygen for use by active (hence carbon dioxide producing) tissues. After Christian Bohr, 1855–1911, Danish physiologist.

Bradycardia: Unusually slow heartbeat, usually considered as under 60 per minute.

Bronchi: The branching pipes of the lungs connecting the main *tracheal* pipes from the head with the *alveoli.*

Bundle of His: A set of modified heart muscle fibers that run from the *atrioventricular node* out to the various parts of the ventricles; they convey the signals for ventricular contraction and ensure that the whole of the pair of *ventricles* squeezes at once. After Wilhelm His, 1863–1934, Swiss cardiologist.

Bypass surgery: Connection of a set of pipes from the *aorta* around past the main parts of one or more of the *coronary arteries.* The pipes are usually *veins* taken from the upper legs where several run parallel; the operation circumvents areas of major blockage of the coronary arteries and thus improves the blood supply to the heart muscle.

Calorie: a unit of energy or *heat*, defined as the amount of heat needed to raise the *temperature* of 1 gram of water by 1° Celsius. It is confusingly similar to the Calorie (upper case, but inconsistent) used in nutrition, 1000 calories or the amount of heat needed to raise the temperature of a kilogram of water by that 1° Celsius. The Calorie is better termed the kilocalorie. For scientific use, the calorie has largely been replaced by the joule, where 1 cal = 4.2 J, with the joule defined in mechanical rather than thermal terms.

Capillaries: The smallest pipes of the circulatory system, the sites of exchange of material between blood and surrounding tissue, with nonmuscular walls only a single cell thick.

Carbon dioxide: A gas, quite soluble in water, produced by the complete combustion (= combination with *oxygen*) of carbon-containing chemicals whether in an open fire, an internal combustion engine, or an active cell supplied with ample oxygen.

Carbon monoxide: A highly toxic gas produced by the incomplete combustion of carbon-containing chemicals; toxicity results from nearly irreversable binding with

hemoglobin and consequent interference with *oxygen* transport by blood.

Cardiac: Referring to the heart

Cardiac catheterization: Threading a flexible tube through *veins* or *arteries* into the heart or the blood vessels adjacent to it in order to make measurements, inject tracer substances, perform mechanical procedures such as *angioplasty*, and so on.

Cardiac output: The volume per unit time coming out of either right or left *ventricles* (since, in mammals and birds, the two must be the same); equal to stroke volume times heartbeat rate or stroke frequency. The conventional definition is, of course, only a half of the actual output of a heart.

Cephalopods: A group of *mollusks* including squid, octopus, cuttlefish, and nautilus, but excluding snails, clams, scallops, and so on. Literally, "head-footed"—all bear prehensile arms on their heads.

Cerebrospinal fluid: The liquid bathing the central nervous system; similar but not the same in composition as blood *plasma*.

Cholesterol: Strictly, a single chemical of the general group termed "sterols" that includes many hormones; it's essential for making cell membranes and can be made by animals such as we from almost any digestible material. Excessive levels of blood cholesterol, though, are associated with *atherosclerosis* and are commonly correlated with obesity and specific dietary characteristics. In looser usage, the name is given to certain results of standard blood analyses, in which a fraction termed "low density lipoproteins" or LDLs are called "bad cholesterols" and "high density lipoproteins" or HDLs are called "good cholesterols." The explicit value judgments refer to prognostication of *atherosclerosis* and *coronary artery disease.*

Cilia: Actively moving filimentous outgrowths of individual cells; they usually either propel a free-living cell or move liquid or mucous material along a tube or the surface of some organ. Flagella are structurally and functionally similar.

Closed circulatory system: One such as that of *vertebrates* and *cephalopod mollusks* in which the circulating fluid is entirely contained within a set of pipes and pumps. (contra *open c.s.*)

Collagen: A protein used by animals for mechanical purposes; the main constituent of tendons but found also in bone, cartilage, and other places. Insoluble in water unless treated with acid; thereafter the basis of gelatin and various animal-derived glues.

Conductivity, thermal: A measure of how fast *heat* moves from warmer to cooler locations in a material (or from one material to another) by direct transfer of molecular momentum (conduction); no radiation or actual material transfer is involved.

Convection, free and forced: *heat* transfer through relative movement (flow) of warmer and cooler material (by contrast with *conductivity, thermal*). In free convection the movement is driven by gravity and the different densities of warm and cool material (as in the rise of hot air); in forced convection some pump or fan drives the motion.

Coronary arteries: The arteries that supply oxygenated blood to the muscle of the heart, located on the exterior surface of the heart.

Coronary artery disease: *Atherosclerosis* of the *coronary arteries* leading to *ischemia, angina pectoris, infarction,* and other evils.

Corpuscle: Purist name for "cell" in *"red blood cell"* to recognize that mammalian RBC's have no nuclei and thus aren't properly competent cells.

Countercurrent exchanger: Device to transfer *heat* or dissolved material such as *oxygen* from one moving *fluid* to another; the "countercurrent" feature involves the fluids moving in opposite directions and permits very highly efficient or nearly complete transfer.

Diaphragm: A muscular wall, concave upward, that separates the contents of the *thorax* (heart, lungs, etc.) from that of the abdomen (stomach, intestine, liver, kidneys, etc.) in mammals. Downward movement by muscular contraction is one factor driving inhalation of air.

Diastole: The stage of the heartbeat in which the muscle relaxes and the *ventricles* refill.

Diffusion, facilitated diffusion: Movement of one material through another by nothing more than chaotic molecular wandering; in facilitated diffusion a carrier molecule, typically *myoglobin*, enhances the speed of the process.

Dissociation curve: A graph showing the relationship between *hemoglobin*–oxygen association and the local concentration of *oxygen* available for association.

Echocardiography: Diagnostic procedure that constructs an image of heart and associated vessels by exposing them to very high pitched sound and picking up reflected sound; similar to sonar as used underwater.

Edema: Excessive accumulation of watery (*interstitial*) fluid in body tissues, whether local or general.

EKG (sometimes ECG): See *electrocardiogram*.

Elasticity: See *extensibility, stiffness,* and *resilience.*

Elastin: A water-insoluble fibrous protein used for mechanical purposes; found as a component of the walls of blood vessels and elsewhere, most notably in the large ligament running along the back of the neck of grazing mammals such as sheep. It's much more *extensible* than *collagen*.

Electrocardiogram: Recording made from electrodes attached to the skin of electrical signals (voltages) that reflect the events associated with the heartbeat.

Embolism: Obstruction of a blood vessel by some foreign material such as a wedged-in blood clot (a thrombus); real trouble if in *coronary arteries*, lungs, or brain. Also, a froth of air in the heart that prevents the heart, a *liquid* pump, from propelling blood.

Erythrocytes: *red blood cells* or red blood *corpuscles* or RBCs.

Eustachian tubes: Tiny pipes from nasal passages to middle ear; they vent the air space in the middle ear to the atmosphere so average *pressures* are the same on the two sides of the eardrum. They often get plugged with nasal mucus, impairing hearing and causing ear pain if one is exposed to changes in atmospheric pressure.

Extensibility: The degree (fraction of original length) to which a material can be stretched before breakage or damage—irrespective of the force or *stress* needed to do the stretch. It's one of the variables (along with *resilience* and the inverse of *stiffness*) mixed up in the vernacular term "elasticity."

Facilitated diffusion: See *diffusion*.

Fat: Somewhat chemically diverse class of nutritive materials characterized by being insoluble in water, greasy or oily in texture, and yielding substantially more energy upon oxidation (when burned in or out of body) than carbohydrates or proteins. (= lipid)

Feedback, negative: A sample of the output of a machine is detected and sent back as an input to that machine. This feedback input is used to readjust the operation of the machine so as to offset any deviation of the output from some pre-established norm.

Feedback, positive: A sample of the output of a machine is detected and sent back as an input to that machine. The feedback input is used to trigger a change in the operation of the machine that increases any deviation of the output from the previous value.

Fibrillation: Rapid and largely uncoordinated twitching of muscle fibers. *Atrial* fibrillation is a serious but manageable matter; *ventricular* fibrillation is very quickly fatal if not treated since it effectively stops *cardiac output*.

Fibrin: A fibrous protein formed in blood clotting from its unimaginatively named soluble precurser, fibrinogen.

Fibrinogen: See *fibrin*.

Filariasis: Parasitic disease involving blockage of lymphatic vessels and consequent extreme *edema* caused by filarial worms.

Flagella: See *cilia*.

Fluid: A substance that can flow, whether in a *gaseous* or a *liquid* state.

Gas: A state of matter in which the substance has neither intrinsic size nor shape but fills any container in which it's placed (contra, *liquid*). A gas dissolved in a liquid is no longer a gas; however, it's sometimes referred to as a gas for convenience. Thus *oxygen* and *carbon dioxide* exchange across *capillary* walls may be called "gas exchange" even though neither is in the gaseous state at any point in the exchange.

Gauge pressure: *Pressure* above and beyond the prevailing ambient or atmospheric pressure (also, "gage" p, an offensive neologism).

Gill: Organ for *gas* exchange between blood and ambient water, present in fishes, fully aquatic *amphibians*, most crustaceans, aquatic insect larvae, *mollusks*, and many other animals of moderate to large size.

Gradient: The change of some variable quantity across some measureable distance, as in velocity gradient, concentration gradient, or *pressure* gradient.

Hagen–Poiseuille equation: Equation relating the total volume flow through a cylindrical pipe to the *pressure* difference and length of the pipe (the pressure gradient), to the *viscosity* of the flowing *fluid*, and to the diameter of the pipe; works only for *laminar flow*, and several other conditions must be met for strict applicability. After Gotthilf H. L. Hagen, 1797–1884, German engineer, and Jean Louis Poiseuille, 1799–1869, French physiologist and physician.

Heart attack: Sudden inability of the heart to pump properly, commonly associated with *ischemia* caused by an *embolism*—in short, something blocking a *coronary artery*. (= myocardial *infarction*)

Heartburn: Sharp chest pain caused by movement of acidic stomach contents up into the esophagus; no connection with hearts, per se; also the double entendre title of a novel by Nora Ephron. (= pyrosis)

Heat: A form of energy, measured in the units of energy such as calories or joules. Easily confused with *temperature*, the latter essentially a measure of the intensity rather than the amount of heat.

Hematocrit: A measure of the relative volume of the blood occupied by the cells (mainly *red blood cells*), expressed as a percentage.

Hemocyanin: *Respiratory pigment* (*oxygen* carrier) found in the blood of various invertebrates such as many crustaceans and *mollusks*.

Hemoglobin: *Respiratory pigment* (*oxygen* carrier) found in the bloods of a wide diversity of animals including some worms and all people. Present in a few plants and elsewhere, where it functions in facilitation of *diffusion*.

Hemophilia: Hereditary disease involving a defective blood coagulation system and consequent severe problems with *hemorrhages*

Hemorrhage: Bleeding—escape of blood from the circulatory system, whether internal or external.

Hepatic: Referring to the liver.

Hookean material: Any material in which the degree of deformation (usually stretch) is directly proportional to the force or *stress* applied—thus giving a straight line on a *stress*–strain graph. Biological materials are rarely hookean, but metals commonly are. After Robert Hooke, 1635–1703, English polymath.

Hydrostatic support: Arrangement for mechanical support of an organism or part of one in which *stiffness* is provided, not by hard materials but by a core of *liquid* surrounded by a tensile membrane. The liquid core may, as in a human penis, be provided by the blood of the circulatory system.

Hyperlipidemia: Presence of an abnormally large amount of *fat* in the blood; hypercholesterolemia is a little more specific, referring to *cholesterol* in particular. (The "-emia" refers to blood.)

Hypertension: Persistently high

blood *pressure;* a common condition with no obvious symptoms but with an assortment of undesirable consequences.

Hypotension: Persistently low blood *pressure;* various manifestations include dizziness upon standing up and lightheadedness. Can be but isn't always a serious matter.

Infarction: Sudden reduction in blood supply to an area from a variety of causes. See *heart attack.*

Inferior vena cava: The common collecting point for blood from the abdomen and rear appendages of many *vertebrates;* blood from there commonly enters the right *atrium.*

Interstitial fluid: The fluid between the cells of the body, mainly composed of material squeezed (*ultrafiltered*) from blood *plasma* through the *capillary* walls. The source of, and essentially identical to, *lymph.*

Ischemia: Local reduction in blood supply usually due to some vascular obstruction; thus local *anemia.*

Kilocalorie: See *calorie.*

Laminar flow: Nonchaotic flow in which all bits of fluid move in the direction of the flow as a whole—overall flow isn't simply some average over time as in the contrasting form, *turbulent flow.*

Laplace's law: The rule that the *pressure* within a closed vessel is equal to the *tension* in its wall divided by its radius of curvature (for cylinders and thus blood vessels) or equal to twice the tension in its wall divided by its radius of curvature (for spheres and thus, roughly, heart chambers or *alveoli*). After Pierre-Simon, marquis de Laplace, 1749–1827, French mathematician and astronomer.

Leukocytes: White blood cells, as distinguished from *red blood cells;* normally nucleated cells coming in a variety of types and having a variety of functions. Far less numerous than RBCs.

Lipid: See *fat.*

Liquid: A state of matter in which the substance has intrinsic size (volume) but no specific shape and thus rests in the bottom of any container in which it's placed. (contra, *gas.*)

Lymph: *Liquid* returned to the circulatory system from the tissues through the lymphatic vessels. Essentially the same as *interstitial fluid.*

Lymph hearts: Specific propulsive chambers driving *lymph* heartward; present in lower *vertebrates* but not in mammals.

Lymph nodes: Solid, spherical or bean-shaped bodies located along the course of lymphatic vessels; they function as part of the immunological system.

Lymphocytes: A kind of *leukocyte* formed in the *lymph nodes,* in adult humans making up around a quarter of the body's leukocytes.

Manometer: Any of a variety of devices that measure *pressure*, of which *barometers* and *sphygmomanometers* are particular subsets.

Mediastinum: The wall separating the left and right pleural (lung) cavities in the *thoraces* of mammals. In it are located heart, major blood vessels, and esophagus.

Met: See *metabolic scope*.

Metabolic rate: The rate at which an organism uses energy for all of its activities including the production of *heat*. Commonly measured by following *oxygen* consumption, but that misses *anaerobic* metabolism. Sometimes measured by following heat production. Commonly quoted after dividing by body mass; this latter, relative metabolic rate, is properly termed the "specific metabolic rate."

Metabolic scope: The factor by which the *metabolic rate* of a resting animal is raised in maximal sustained activity. (The specific factor is often referred to as the number of mets, as in, say, 15 mets, except in Shea Stadium, where only 9 mets at a time are permitted.)

Metabolic time: A crude time scale reflecting the intensity of life as given by *metabolic rate* per unit mass of animal (specific metabolic rate); thus metabolic time passes more quickly for a small mammal than for a large one. The duration of a unit of metabolic time is inversely proportional to specific metabolic rate.

Millimeters of mercury (mm Hg). Unit of *pressure* representing the height, in millimeters, to which a column of the densest *liquid*, mercury, can be pushed by a given pressure. One atmosphere at sea level equals 760 mm Hg.

Mitral stenosis: Obstruction of flow from left *atrium* to left *ventricle* due to trouble around the *mitral valve* that separates these chambers; *valve* replacement is often done to fix matters.

Mitral valve: The one-way *valve* between left *atrium* and left *ventricle*, permitting flow only from former to latter.

Mollusks: One of the three major groups of really complex animals, the other two being *vertebrates* and *arthropods;* includes snaillike and clamlike forms, as well as the *cephalopods*. All have soft bodies with or without a hard shell or internal bones.

Mono-unsaturated fat: A lipid of the general sort called *"triglycerides,"* specifically one with a single double chemical bond in each long chain. Olive and canola oils (both mixtures of chemicals) contain especially large amounts of mono-unsaturated fat. (comp. *saturated fat, polyunsaturated fat*)

Murmur: An unusual sound produced by flow through the heart signalling some abnormality in the pattern of flow.

Murray's law: The rule that for minimum cost of *fluid* transport in a branching system of circular pipes with *laminar flow*, the radii of the pipes at branches should have the following characteristic: the cube of the radius of the original pipe should equal the sum of the cubes of the radii of the branches from it. After Cecil D. Murray, 1897–1935, American physiologist.

Muscle: A tissue that is capable of shortening against a load, in the process converting chemical to mechanical energy. Sometimes (skeletal muscle) external stimuli are needed to initiate contraction; for other muscles (*cardiac,* smooth or visceral) intrinsic activity is sufficient.

Myo-: Referring to *muscle*.

Myocardium: The *muscle* of the heart.

Myogenic: Hearts, such as those of *vertebrates,* that beat without requiring external nervous stimuli; external stimuli, however, are used to speed up or slow down the heartbeat. (contra *neurogenic*)

Myoglobin: Muscle *hemoglobin*—a version of hemoglobin present in active, *aerobic* muscles (and some other places) that has the effect of speeding ("facilitating") *diffusion* of *oxygen* from blood to active tissue. It's what makes dark meat dark.

Neurogenic: Hearts, such as are found in some invertebrates, that require a steady input of nerve impulses or they won't beat. (contra *myogenic*)

No-slip condition: The inevitable situation wherever a fluid flows across a solid surface: at the surface, the speed of flow is zero, so a velocity gradient from zero speed to the speed of the flow far from the surface occurs entirely within the fluid.

Open circulatory system: One such as that of *arthropods* and some non-*cephalopod mollusks* in which the circulating *fluid* for part of its circuit flows through what are called blood sinuses, roughly the same as *interstitial spaces.* (contra *closed c.s.*)

Oxygen: Atmospheric component (about 21 percent) capable of combining with chemicals containing carbon and hydrogen and thereby liberating energy, *carbon dioxide,* and water. That energy is the basic power source for the chemical, mechanical, and thermal activities of most animals.

Pacemaker: The point of initiation of the rhythm of the heartbeat, normally the *sinoatrial node;* also any prosthetic electronic device (external or internal) that sends signals that replace a nonfunctioning sinoatrial node.

Papillary muscles: Muscles within the heart that run from the ventricular walls up toward the atrioventricular *valves;* they connect via tendons with the latter and seem to help keep the valves from opening in the wrong direction.

Parabola: A curve on a plane surface, sharply curved in one place and then becoming decreasingly curved on either side.

Parabolic flow: The so-called flow profile for *laminar flow* far from the entrance in a straight, cylindrical pipe. Maximum speeds are around the center of the pipe and the speed tapers to zero at either opposite wall; the distribution of speeds across a diameter of the pipe is *parabolic*.

Pascal: The unit of *pressure* in modern scientific practice; it's equal to one newton per square meter or about a hundred-thousandth of sea level atmospheric pressure. After Blaise Pascal, 1623–1662, French mathematician, physicist, and computer-builder.

Pericardial cavity: The fluid-filled space in which the heart lies, its inner and outer surfaces lined with the pericardial membranes.

Pericarditis: Inflammation of the membranes lining the *pericardial cavity*, commonly from infection. The discomfort mimics a heart attack, but, unlike that of a heart attack, it's aggravated by motion of the *thorax*.

Peristalsis: Progressive contraction and relaxation of the muscular wall of a tube such as the intestine that results in pushing a *liquid* or slurry through the tube.

Phlebotomy: Therapeutic procedure in which a person is relieved of a significant amount of blood; formerly widely used on people from Robin Hood to George Washington, it turns out to be of demonstrable value only for a few uncommon conditions.

Piloerection: Making the hair stand up away from the skin by contraction of minute muscles called the "erector pili"; important in mammalian thermoregulation although a bit of a bad joke in humans.

Plaque: A deposit on a surface including disparate items such as dental plaques on teeth and atheromatous (see *atherosclerosis*) plaques of fatty material on the inner walls of blood vessels.

Plasma: The blood minus its suspended cells (red and white blood cells, mainly), but including all else.

Platelets: Disk-shaped cellular (but nonnucleated) elements in blood, functioning in the initial stage of blood clotting.

Polyunsaturated fat: A lipid of the general sort called "*triglycerides*," specifically one with more than one double chemical bond in each long chain. Safflower oil (a mixture of chemicals) contains especially large amounts of polyunsaturated fat. (comp. *saturated fat, mono-unsaturated fat*)

Portal system: A part of a venous system in which venous blood, instead of returning in ever-larger vessels to the heart, goes into a set of re-ramifying vessels (beginning with a "portal vein") and eventually into another set of *capillaries*. Our main one is the *hepatic* portal system ending in the liver; the name traces back to the ancient Greek physician and anatomist, Galen.

Premature beats: Several kinds of *arrhythmia*, including *atrial* premature beats and *ventricular* premature beats. Either of these may be essentially innocent, associated with heart disease, or a side effect of drugs. The sensation of either may be of a skipped beat, a flutter, or of extra beats.

Pressure: A force exerted over an area, measured as force per unit area, most commonly exerted by a contained body of *fluid* on the walls of the container. By contrast with *stress*, pressure is omnidirectional—the orientation of the wall with respect to the body of fluid is inconsequential except that the fluid exerting the pressure must be on one side only.

Principle of continuity: The rule that for a *fluid* flowing along a pipe the result of multiplying the speed of flow times the cross-sectional area of the pipe will be the same anywhere along the pipe. If the pipe branches, the sum of those multiplications for the daughter pipes will equal the value for the parental pipe.

Pulmonary: Referring to lungs.

Pulmonary circuit: In *vertebrates* with essentially divided circulations (reptiles, birds, mammals), the circuit from heart through lungs (as opposed to elsewhere) and back to heart. (contra *systemic circuit*)

Pulmonary trunk: The great vessel emerging from the mammalian right *ventricle,* analogous to the *aorta* from the left; it immediately splits into right and left *pulmonary* arteries.

Pulse: The dilation of any of the larger arteries that follows ejection of blood from the heart and thus marks the *systolic* phase of the heartbeat. It can be felt by touching or lightly pressing the skin wherever an *artery* runs beneath.

Purkinje fibers: Branches of the *bundle of His* from the *atrioventricular node,* running out over the inner surfaces of the *ventricles* as part of the triggering system. After Jan Evangelista Purkinje, 1787–1869, Czech physiologist.

Random walk: A mathematically idealized progression in which the direction of each step of the progression is randomly chosen and totally independent of the direction of preceding and succeeding steps.

Ratio: The result of dividing one number by another. If the numbers represent physical quantities, they are often expressed in the same dimensions and units; thus one force may be divided by another with a result that's a pure number, with neither dimensions nor units. The ratio called the *Reynolds number* is such a dimensionless pure number.

Red blood cells (or "*erythrocytes*", or "RBCs"): The containers for *hemoglobin* in *vertebrate* bloods; they commonly make up almost half the volume of the blood. In mammals, but not in other vertebrates, they lack nuclei at functional maturity. They consequently can't reproduce themselves, and they have a fairly short useful life.

Reflux: A backward flow, usually referring to gastroesophageal reflux, in which stomach contents move back into the esophagus and produce the sensation called *heartburn*.

Renal: Referring to kidneys.

Resilience: The degree to which a deformed material will return to its original shape when the deforming force is removed. It's one of the variables (along with *extensibility* and the inverse of *stiffness*) mixed up in the vernacular term "elasticity."

Respiratory pigment: Any bloodborne chemical that combines loosely enough with *oxygen* so it will take up oxygen at a site of relatively high availability (for in-

stance, lungs, gills, or skin) and release oxygen at sites of lower availability, especially metabolically active tissues.

Rete mirabile: Literally "wonderful net"—an interwoven mesh of tiny *arteries* and *veins* in sufficiently intimate contact so that either *heat* or a substance such as *oxygen* will easily *diffuse* from one kind of vessel into the other kind; associated (I think inevitably) with *countercurrent exchangers*.

Reynolds number: In fluid mechanics, the *ratio* of so-called inertial forces to *viscous* forces— loosely of those forces keeping a fluid flowing to those slowing it down; its value is an index to the general character of flow, to whether, for instance, flow will be *laminar* or *turbulent*. After Osborne Reynolds, 1842–1912, British engineer and physicist.

Sclerosis: A hardening of a tissue, commonly but not limited to addition of fibrous material.

Serum: Blood *plasma* after the removal of the *fibrinogen* by filtering out clotted material.

Shear: Motion in which two parallel surfaces move past each other; the surfaces may be just arbitrarily chosen planes within, say, a flowing *fluid* or a solid object; the motion typically involves deformation of a material. Complements bending and twisting deformations.

Sinoatrial node (S-A node): An area of specialized heart tissue at which a nerve enters the heart and at which the heatbeat begins; thus the normal, intrinsic *pacemaker* of the heart. Located in the wall of the right *atrium* near its junction with the *superior vena cava*.

Skimming flow: Flow into a branching vessel from a main one in which the branching vessel, coming out from the wall of the main one, receives blood from the relatively cell-free layer near the wall. Thus it gets relatively more *plasma* and fewer cells than are present in normal blood.

Sphygmomanometer: A *manometer* and accessories for measuring blood *pressure*.

Spleen: A fairly large organ lying in the abdomen between stomach and *diaphragm* that acts as a culling organ for aged or imperfect *red blood cells*, as a reservoir for *platelets*, and a producer of *lymphocytes*. No function is both unique to it and critical for survival.

Starling hypothesis: The mechanism for using blood pressure and osmotic pressure gradients from *arteriolar* to *venular* ends of a *capillary* to exchange material across the capillary wall, pushing out fluid and dissolved molecules at the arteriolar end and drawing them back in at the venular end. After Ernest H. Starling, 1866–1927, British physiologist.

Starling's law: The idea, stated by Ernest H. Starling in 1918, that the force of contraction of the heart is proportional to the extent to which the heart *muscle* is initially stretched. Also referred to as the Frank–Starling mechanism, to recognize Otto Frank (see *Windkessel*), who had somewhat the same idea.

Stenosis: A narrowing of a channel; usually referring to any of the orifices containing the various heart *valves*.

Sternum: The stiff front of the chest, made up of cartilage connecting the otherwise free ends of the ribs.

Stethoscope: An instrument used to listen to the various sounds produced by the operation of the body's organs—especially lungs, heart, and digestive system.

Stiffness: The resistance of a material to deformation; thus the deforming *stress* divided by the extent of deformation (*strain*) caused by the stress. Its inverse (strain divided by stress) corresponds to one of the variables (along with *extensibility* and *resilience*) mixed up in the vernacular term "elasticity."

Strain: The relative extent of a deformation, commonly given as the change in length of the object as a result of the deforming force divided by the object's original length. Distinct from *stress*.

Stress (1): The force applied to an

object in an attempt to deform it (rather than move it, usually) divided by the area over which the force is distributed. Similar to pressure, but unidirectional. Distinct from *strain*.

Stress (2): Psychological tension or strain attributed to personality, life-style, and so on. Stress is supposedly not good for cardiovascular health; but, as common with mental matters, trustworthy data is confounded by arbitrariness in the operational definition of the variable at issue. No measurement, of course, can be better than the definition of what's measured!

Stroke volume: The volume that comes out of either *ventricle* in a single heartbeat; if multiplied by rate of heartbeat, it yields *cardiac output*. This is, of course, half of the total output of the heart as a dual pump.

Supercharger: A device that forces more air or combustible mixture into an engine than the engine would normally draw in, thus increasing the power the engine can develop.

Superior vena cava: The common collecting point for blood from the head, upper body, and front appendages of many *vertebrates;* blood from there commonly enters the right *atrium*.

Syncope: Fainting, temporary loss of consciousness as a result of deficiency of blood flow to the brain.

Systemic circuit: In vertebrates with essentially divided circulations (reptiles, birds, mammals), the circuit from heart through body (as opposed to lungs) and back to heart. (contra *pulmonary circuit*)

Systole: The squeezing phase of the heartbeat in which *ventricular* contraction forces blood into the arterial system.

Tachycardia: Abnormally rapid heartbeat, usually applied to resting rates over 100 per second. (contra *bradycardia*.)

Temperature: The degree of hotness or coldness of anything, commonly measured in degrees Celsius or degrees Fahrenheit; often confused with *heat*, a form of energy.

Tension: The force tending to stretch something such as a membrane divided by the distance across which the force is exerted. Slightly different from both tensile force (a force, *per se*) and tensile stress (a force divided by an area).

Thorax: The portion of the trunk of a mammal in front of (above or anterior to) the *diaphragm;* it contains lungs, heart, esophagus, ribs, and so on.

Torpor: A reduction in *metabolic rate*, body *temperature*, heartbeat, and other functions in a normally warm-blooded animal; it may be associated with hibernation, although in some very small forms nightly torpor occurs.

Trabeculae: Supporting fibers running across a structure; in the heart the "trabeculae carneae" (usually including the *papillary muscles*) extend inward in ridges or columns from the inner *ventricular* walls.

Tracheae: In terrestrial *vertebrates* the windpipe, extending from head into the *thorax* where it branches into the main *bronchi*. In insects the pipes of the system that distributes air directly to the metabolically active tissues.

Tricuspid valve: The one-way *valve* between right *atrium* and right *ventricle,* contrasting with the bicuspid (two- rather than three-element) or *mitral valve* between left atrium and ventricle.

Triglycerides: *Fat* molecules of the most ordinary sort, each formed of one glycerol molecule and three fatty acids. The latter are a diverse lot and include *saturated, monounsaturated,* and *polyunsaturated* versions.

Turbulent flow: Chaotic flow in which bits of *fluid* move in separate directions and the direction of the flow as a whole even at a single point is no more than an average over time, in contrast with nonchaotic *laminar flow.*

Ultrafiltration: Filtration of *liquid* and dissolved molecules (but not cells and very large molecules)

through a membrane, driven by hydrostatic *pressure* as in the initial stage of a *vertebrate* kidney or across *capillary* walls according to the *Starling hypothesis.*

Valsalva maneuver: Forced exhalation against a closed nose and mouth either to open the *eustachian tubes* or to increase intrathoracic pressure. The latter reduces venous return to the right *atrium,* flow through the lungs, *cardiac output,* and *coronary* blood flow. After Antonio Maria Valsalva, 1666–1723, Italian anatomist.

Valve: A device that controls the passage of material through a channel; often control is simply of direction, with passage permitted in one direction but not the other—as in those of heart and veins.

Vasoconstriction: Short-term narrowing of small blood vessels, usually through contraction of *muscles* in their walls.

Vasodilation: Short-term widening of small blood vessels, usually through relaxation of *muscles* in their walls.

Vein: Vessels conveying blood toward the heart, including the *pulmonary* veins and the *venules.*

Vena cava: See *superior vena cava, inferior vena cava.*

Venous valves: Small one-way *valves* in the *veins* of the extremities that prevent flow away from the heart;

with local muscular action they form an auxiliary heartward pumping system as well. Discovered by Fabricius (Girolamo Fabrizio), 1537–1619, Italian anatomist and teacher of Harvey.

Ventricle: The main, highly muscular, pumping chamber of a heart; two in number in birds and mammals, one (variously partitioned) in lower *vertebrates*.

Venules: The smallest of the *veins*, only a little larger than *capillaries*, but with muscular walls.

Vertebrates: One of the three major groups of really complex animals, the other two being *mollusks* and *arthropods;* includes fishes, *amphibians*, reptiles, mammals, and birds. All have a vertebral column, bones that grow, and some other specific features.

Viscosity: The resistance to flow of a *fluid*, whether *gas* or *liquid*. Quite distinct from *stiffness*, the resistance to deformation, which is a variable that characterizes solids.

Windkessel: A model for how the *extensibility* and *resilience* of arterial walls reduces the extreme blood *pressure* fluctuations caused by the pulsatory character of the heart as a pump, devised by F.W.F. Otto Frank, 1865–1944, German physiologist.

References

Agnisola, C. 1990. Functional morphology of the coronary supply of the systemic heart of *Octopus vulgaris*. *Physiol. Zool.* 63: 3–11.

Bartlett, J. 1955. *Familiar Quotations*. Boston, MA: Little, Brown & Co. 1614 pp.

Beard, J. A. 1970. *The James Beard Cookbook*. New York: E.P. Dutton & Co. 544 pp.

Benzinger, T. H. 1964. The thermal homeostasis of man. In C. F. A. Pantin, ed.: *Homeostasis and Feedback Mechanisms*. Symp. Soc. Exp. Biol. Vol. 18. New York: Academic Press, pp. 49–80.

Berkow, R. 1987. *The Merck Manual of Diagnosis and Therapy, 15th ed.* Rahway, NJ: Merck Sharpe & Dohme Research Laboratories. 2165 pp.

Bernard, C. 1865. *An Introduction to the Study of Experimental Medicine.* reprint, 1957. New York: Dover Publications, Inc. 266 pp.

Beveridge, W. I. B. 1957. *The Art of Scientific Investigation,* 3rd ed. New York: W. W. Norton. 177 pp.

Bogen, D. K. and T. A. McMahon. 1979. Do cardiac aneurysms blow out? *Biophys. J.* 27: 301–16.

Bourne, G. B. 1987. Hemodynamics in squid. *Experientia 43*: 500–502.

Bramble, D. 1989. Axial-appendicular dynamics and the integration of breathing and gait in mammals. *Amer. Zool.* 29:171–86.

Burns, J. M. 1975. *Biograffiti*. New York: Quadrangle/New York Times Book Co. 112 pp.

Burton, A. C. 1972. *Physiology and Biophysics of the Circulation,* 2nd ed. Chicago: Year Book Medical Publishers, Inc. 226 pp.

Calder, W. A. 1984. *Size, Function, and Life History*. Cambridge, MA: Harvard University Press. 431 pp.

Cannon, W. 1932. *The Wisdom of the Body*. New York: Norton. 312 pp.

Caro, C. G., T. J. Pedley, R. C. Schroter, and W. A. Seed. 1978. *The Mechanics of the Circulation*. Oxford: Oxford University Press. 527 pp.

Chien, S., S. Usami, R. J. Dellenback, and M. I. Gregersen. 1967. Blood viscosity: influence of erythrocyte deformation. *Science* 157: 829–31.

Doyle, Sir A. Conan. 1887. *A Study in Scarlet*. London: Ward Lock & Co. 144 pp.

Doyle, M. P., W. R. Galey, and B. R. Walker. 1989. Reduced erythrocyte deformability alters pulmonary hemodynamics. *J. Appl. Physiol.* 67: 2593–99.

Edholm, O. G. 1978. *Man—Hot and Cold*. London: Edward Arnold. 60 pp.

Folkow, B., and E. Neil. 1971. *Circulation*. New York: Oxford University Press. 593 pp.

Frank, R. G., Jr. 1980. *Harvey and the Oxford Physiologists*. Berkeley, CA: University of California Press. 368 pp.

Fung, Y. C. 1984. *Biodynamics: Circulation*. New York: Springer-Verlag. 404 pp.

Gosline, J. M., and M. E. DeMont. 1985. Jet-propelled swimming in squids. *Sci. Amer.* 252(1): 96–103.

Gould, S. J. 1980. *The Panda's Thumb*. New York: W.W. Norton & Co. 343 pp.

Graubard, M. 1964. *Circulation and Respiration: The Evolution of an Idea*. New York: Harcourt, Brace, & World. 278 pp.

Gregg, D. E., and L. C. Fisher. 1963. Blood supply to the heart. In W. R. Hamilton and P. Dow, eds.: *Handbook of Physiology. Sect. 2: Circulation, Vol. 2*. Washington, DC: American Physiological Society, pp. 1517–84.

Harvey, W. 1628. *On the Motion of the Heart and Blood in Animals*. (Willis's translation, 1962. Chicago: Henry Regnery Co., 213 pp. with introduction and *Two Letters Addressed to John Riolan*.)

Hyman, L. H. 1942. *Comparative Vertebrate Anatomy, 2nd ed*. Chicago: University of Chicago Press. 544 pp.

Horrobin, D. F. 1970. *Principles of Biological Control*. Aylesbury, UK: Medical and Technical Publishing Co, Ltd. 70 pp.

Johansen, K., and A. W. Martin. 1965. Circulation in a giant earthworm, *Glossoscolex giganteus*. I. Contractile processes and pressure gradients in the large vessels. *J. Exp. Biol.* 43: 333–47.

Kaufman, W. I., and S. Lakshmanan. 1964. *The Art of India's Cookery*. Garden City, New York: Doubleday & Co. 239 pp.

Kilgour, F. G. 1952. William Harvey. *Sci. Amer.* 186 (6): 56–62.

Krogh, A. 1941. *The Comparative Physiology of Respiratory Mechanisms*. Philadelphia, PA: University of Pennsylvania Press. 172 pp.

Krogh, A., and P. B. Rehberg. 1924. Kinematographic methods in the study of capillary circulation. *Amer. J. Physiol.* 68: 153–60.

LaBarbera, M. 1990. Principles of design of fluid transport systems in zoology. *Science* 249: 992–1000.

LaBarbera, M., and S. Vogel. 1982. The design of fluid transport systems in organisms. *Amer. Sci.* 70: 54–60.

Langille, B. L., and F. O'Donnell. 1986. Reductions in arterial diameter produced by chronic decreases in blood flow are endothelium-dependent. *Science* 231: 405–7.

Lansman, J. B. 1988. Endothelial mechanoreceptors: going with the flow. *Nature* 331: 481–82.

Latham, R., and W. Matthews. 1976. *The Diary of Samuel Pepys: A New and Complete Translation.* Berkeley, CA: University of California Press (9 vols. + index).

Lillywhite, H. B. 1987. Circulatory adaptations of snakes to gravity. *Amer. Zool.* 27: 81–95.

Loewi, O. 1953. *From the Workshop of Discoveries.* Lawrence, KA: University of Kansas Press. 62 pp.

Machin, K. E. 1964. Feedback theory and its application to biological systems. In C. F. A. Pantin, ed.: *Homeostasis and Feedback Mechanisms.* Symp. Soc. Exp. Biol. Vol. 18. New York: Academic Press, pp. 421–45.

Malpighi, M. 1661. *De Pulmonibus.* translated parts in Graubard, op. cit.

Martin, A. W. 1980. Some invertebrate myogenic hearts: the hearts of worms and molluscs. In G. H. Bourne, ed.: *Hearts and Heart-Like Organs,* Vol. 1. New York: Academic Press, pp, 1–39.

Mayerson, H. S. 1963. The lymphatic system. *Sci. Amer.* 208(6): 80–90.

Mayr, O. 1970. *The Origins of Feedback Control.* Cambridge, MA: M.I.T. Press. 151 pp.

McMahon, T. A. 1987. *Loving Little Egypt.* New York: Viking Penguin, Inc. 273 pp.

McMahon, T. A., and J. T. Bonner. 1983. *On Size and Life.* New York: Scientific American Books. 255 pp.

Meyhöfer, E. 1985. Comparative pumping rates in suspension-feeding bivalves. *Marine Biology* 85: 137–42.

Milnor, W. R. 1990. *Cardiovascular Physiology.* New York: Oxford University Press. 501 pp.

Milsum, J. H., ed. 1966. *Positive Feedback: A General Systems Approach to Positive/Negative Feedback and Mutual Causality.* Oxford: Pergamon Press. 169pp.

Murray, C. D. 1926. The physiological principle of minimum work. I. The vascular system and the cost of blood volume. *Proc. Nat. Acad. Sci. USA* 12: 207–14.

Nordenskiöld, E. 1928. *The History of Biology.* New York: Alfred A. Knopf, Inc. 629 pp.

O'Regan, R. G., and S. Majcherczyk. 1982. Role of peripheral chemoreceptors and central chemosensitivity in the regulation of respiration and circulation. *J. Exp. Biol.* 100: 23–40.

Ottaviani, G., and A. Tazzi. 1977. The lymphatic system. In Gans, C., ed.: *Biology of the Reptilia. Vol 6.* London: Academic Press, pp. 315–462.

Parry, D. A., and R. H. J. Brown. 1959. The hydraulic mechanism of the spider leg. *J. Exp. Biol.* 36: 423–33.

Perkins, W. L. 1965. *The Fanny Farmer Cookbook.* Boston: Little, Brown & Co. 624 pp.

Prosser, C. L. 1973. *Comparative Animal Physiology.* Philadelphia: W. B. Saunders Company. 966 pp.

Reynolds, O. 1883. An experimental investigation of the circumstances which determine whether the motion of water shall be direct or sinuous, and of the law of resistance in parallel channels. *Trans. Roy. Soc. Lond.* 174: 935–82.

Robinson, T. F., S. M. Factor, and E. H. Sonnenblick. 1986. The heart as a suction pump. *Sci. Amer.* 254(6): 84–91.

Roddie, I. C. 1990. Lymph transport mechanisms in peripheral lymphatics. *News in Physiol. Sci.* 5: 85–89.

Root-Bernstein, R. S. 1989. *Discovering.* Cambridge, MA: Harvard University Press. 501 pp.

Rouse, H., and S. Ince. 1957. *History of Hydraulics.* reprint, 1963. New York: Dover Publications, Inc. 269 pp.

Ruud, J. T. 1965. The ice fish. *Sci. Amer.* 213(5): 108–14.

Schmidt-Nielsen, K. 1972. *How Animals Work.* Cambridge: Cambridge University Press. 114 pp.

Schmidt-Nielsen, K. 1981. Countercurrent systems in animals. *Sci. Amer.* 244(5): 118–28.

Schmidt-Nielsen, K. 1984. *Scaling: Why is Animal Size so Important?* Cambridge: Cambridge University Press. 241 pp.

Schmidt-Nielsen, K. 1990. *Animal Physiology: Adaptation and Environment.* Cambridge: Cambridge University Press. 602 pp.

Scholander, P. F. 1957. The Wonderful Net. *Sci. Amer.* 196(4): 96–107.

Scholander, P. F. 1960. Oxygen transport through hemoglobin solutions. *Science* 131: 585–90.

Scholander, P. F. 1990. *Enjoying a Life in Science.* Fairbanks: University of Alaska Press. 226 pp.

Seymour, R. S. 1987. Scaling of cardiovascular physiology in snakes. *Amer. Zool.* 27: 97–109.

Shadwick, R. E., and J. M. Gosline. 1981. Elastic arteries in invertebrates: mechanics of the octopus aorta. *Science* 213: 759–61.

Shadwick, R. E., C. M. Pollock, and S. A. Stricker. 1990. Structure and biomechanical properties of crustacean blood vessels. *Physiol. Zool.* 63: 90–101.

Sharpey-Schafer, E. P. 1955. Effects of Valsalva's manoeuver on the normal and failing circulation. *Brit. Med. J.* 1955(1): 693–95.

Smith, K. K., and W. M. Kier. 1989. Trunks, tongues, and tentacles: moving with skeletons of muscle. *Amer. Sci.* 77: 28–35.

Taylor, C. R., and E. R. Weibel. 1981. Design of the mammalian respiratory system. I. Problem and strategy. *Resp. Physiol.* 44:1–10.

Trollope, A. 1871. *Ralph the Heir.* London: Strahan & Co. 434 pp.

Vander, A. J., J. H. Sherman, and D. S. Luciano. 1985. *Human Physiology: The Mechanisms of Body Function.* New York: McGraw Hill Book Co. 715 pp.

Vogel, S. 1978. Organisms that capture currents. *Sci. Amer.* 239 (2): 128–39.

Vogel, S. 1981. *Life in Moving Fluids.* Princeton, NJ: Princeton University Press. 352 pp.

Vogel, S. 1988. How organisms use flow-induced pressures. *Amer. Sci.* 76: 28–34.

Vogel, S. 1988. *Life's Devices: The Physical World of Animals and Plants.* Princeton, NJ: Princeton University Press. 367 pp.

Voranen, M. 1989. Basic functional properties of the cardiac muscle of the common shrew (*Sorex ananeus*) and some other small mammals. *J. Exp. Biol.* 145: 339–51.

Warren, J. V. 1974. The physiology of the giraffe. *Sci. Amer.* 231(5): 96–105.

Warwick, R., and P. L. Williams. 1973. *Gray's Anatomy: 35th British Edition.* Philadelphia: W.B. Saunders Co. 1471 pp.

Watt, B. K., and A. L. Merrill. 1963. *Composition of Foods.* Washington, D.C.: U.S. Government Printing Office. 190 pp.

Wells, M. J. 1987. The performance of the octopus circulatory system: a triumph of engineering over design. *Experientia* 43: 487–99.

West, J. B. 1982. Respiratory and circulatory control at high altitudes. *J. Exp. Biol.* 100: 147–57.

Wilson, E. B., Jr. 1952. *An Introduction to Scientific Research.* New York: McGraw-Hill Book Company. 375 pp.

Wrigley, E. A. 1969. *Population and History.* New York: McGraw-Hill Book Company. 256 pp.

Zweifach, B. W. 1959. The microcirculation of the blood. *Sci. Amer.* 200(1): 54–60.

Index

Page numbers for definitions are in italics

and lift, 149
in lymphatic vessels, 206
vs. manometric height, 65–68
measuring, 60–61
omnidirectional character, 60
units, 60–62
Primitive vs. advanced, 78, 249, 262–63
Principle of continuity, 75, 78–80, 89, 106, 138, 283
Prostheses
heart valves, 26, 33
pacemaker, 54
Proteins (blood proteins)
in blood, 15
in coagulation process, 240
in lymph, 206–7
release from liver, 225
resistance to filtration, 204, 243
stability vs. temperature, 179
synthesis in liver, 245
urea production from, 242
Pulmonary circuit, 30, 46, 76, 283
mixing with systemic, 83–85
pressures in, 70
Pulmonary trunk and veins, 283
changes at birth, 36
of fetus, 35
and Harvey, 45, 50
location, 28, 29, 31
pressure in, 70
relative size, 75
Pulmonary valve, 32
Pulse rate. *See* Heartbeat rate
Pumps. *See also* Macropumps; Micropumps
biological pumps, kinds, 135
evaporative, 141
as furnaces, 9, 11
heart, 10, 11
macro- and micropumps, 135–41
peristaltic pumps, 135
piston, 148

role, energy supplying, 13
role, pressure imparting, 13–14
supercharging, 23–24, 141

Random walk, 99–100, 283
Red blood cells, 284
amphibians, 196
axial accumulation, 201–2
and blood viscosity, 86, 199–204
in capillaries, 196–97, 203–4
corpuscles, 172
destroyed in spleen, 245
at high altitude, 164
hematocrit, human, 15–16, 164
lifespan, human, 172
in lymph, 206
made in bone marrow, 245
nucleated vs. not so, 172
rotation in small vessels, 201–2
shape, 201, 203
size, 110, 196
speed of travel, 202
total number, human, 16
Rehabilitation, cardiac, 5, 266
Renal portal system, 247–49
Reptiles. *See also* Crocodilians; Snakes
blood pressure, 67
circulatory circuits, 14, 83–84, 183
heart, 83–84
lymph hearts, 208
renal portal systems, 247–49
thermoregulation, 178
Resilience, 284
arterial wall, 118
collagen, 126
elastin, 125
energy conservation, 117
spider silk, 117
Respiratory pigments, 158–60, 284. *See also* Hemocyanin; Hemoglobin